Longman Mathematical Texts

Waves

A mathematical approach to
the common types of wave motion

Longman Mathematical Texts

Edited by Alan Jeffrey and Iain Adamson

Longman Mathematical Texts

Waves
A mathematical approach to
the common types of wave motion

Second edition

The late **C. A. Coulson**
revised by **Alan Jeffrey**

Professor of Engineering Mathematics
in the University of Newcastle upon Tyne

Longman
London and New York

Longman Group Limited
Longman House, Burnt Mill,
Harlow, Essex CM20 2JE England
Associated companies throughout the world

Published in the United States of America
by Longman Inc., New York

© C. A. Coulson
This edition © Longman Group Limited 1977

First published by Oliver and Boyd Ltd. 1941
Published as second edition by Longman Group Ltd. 1977
Reprinted 1981, 1986

Library of Congress Cataloging in Publication Data
Coulson, Charles Alfred.
  Waves.

  (Longman mathematical texts)
  Includes index.
  1. Wave-motion, Theory of. I. Jefffrey, Alan.
II. Title.
QA927.C65 1977 531′.1133′01515 77-3001

ISBN 0-582-44954-5

Set in Linotron Times
Produced by Longman Group (FE) Ltd
Printed in Hong Kong

Preface to the second edition

The book written in 1941 by the late Charles Coulson served to introduce a great many students to the basic ideas that underlie linear wave motion. Just prior to his death he was beginning to give thought to the task of revising his book in order to take account of the changes in the teaching of the subject that have inevitably taken place since it was first written. It is to be regretted that he left no plan indicating the modifications he intended to make. Consequently, when accepting his last request to me to see that a new edition of his book should appear in print, I had to undertake my own re-planning of this work without benefiting from any of his ideas or the special gift of physical insight that he brought to his teaching.

The task of revision is never a straightforward one, and when working on the present book I was always conscious of the fact that a text like "Waves", which has stood the test of time so well, obviously had a content and style of presentation that suited it to the needs of a very wide readership. Accordingly, I considered it proper to preserve as much as possible of the original structure and writing. So, within the framework of the eight chapters comprising the original book, I have confined my own contributions very largely to the inclusion of new sections and examples wherever these seemed to be needed.

In recent years there has been a considerable growth in the study of nonlinear wave phenomena and it seemed important to me that some account of these matters should be given in this new edition. In order to accomplish my task within a modest number of pages, and to keep this first encounter with the subject at a level which is compatible with the rest of the book, rather careful selection of material was necessary. Essentially, this amounted to confining the illustrations used to fluid mechanics, since only there is the structure of problems sufficiently simple to enable them to be formulated without undue time first being spent developing the physical background. Despite this restriction, the chapter contains enough basic material to enable students familiar with more than a first course in subjects as diverse as solid mechanics, plasticity, ferromagnetism, hydrology and chromatography to proceed as far as formulating and solving certain rather simple problems.

In his Preface to the first edition Charles Coulson wrote "the object of this book is to consider from an elementary standpoint as many different types of wave motion as possible." It is my hope that by revising and extending his work as I have, this same statement is as true now as it was when he wrote these words in 1941.

Alan Jeffrey
January 1977

Contents

9: Nonlinear waves

The wave equation

§1   Introduction

We are all familiar with the idea of a wave; thus, when a pebble is
dropped into a pond, water waves travel radially outwards; when a
piano is played, the wires vibrate and sound waves spread through the
room; when a television station is transmitting, electric waves move
through space. These are all examples of wave motion, and they have
two important properties in common: firstly, energy is propagated to
distant points; and, secondly, the disturbance travels through the
medium without giving the medium as a whole any permanent displacement. Thus the ripples spread outwards over a pond carrying
energy with them, but as we can see by watching the motion of a small
floating body, the water of the pond itself does not move in the
direction of propagation of the waves. In the following chapters we
shall find that whatever the nature of the medium which transmits the
waves, whether it be air, a stretched string, a liquid, an electric cable or
some other medium, these two properties which are common to all
these types of wave motion, will enable us to relate them together. In
the linear case they are all governed by a certain differential equation
called the **wave equation** or, as it is sometimes termed, the **equation of
wave motion** (see §5), and the mathematical part of each separate
problem merely consists in solving this equation with the right boundary conditions, and then interpreting the solution appropriately.

§2   General form of progressive waves

Consider a disturbance ϕ which is propagated in the positive direction
along the x axis with velocity c. There is no need to state explicitly what
ϕ refers to; it may be the elevation of a water wave or the magnitude of
a fluctuating electric field. Then, since the disturbance is moving, ϕ will
depend on x and t. When $t = 0$, ϕ will be some function of x which we
may call $f(x)$. $f(x)$ is the **wave profile**, since if we plot the disturbance ϕ

against x, and "photograph" the wave at $t = 0$, the curve obtained will be $\phi = f(x)$. If we suppose that the wave is propagated without change of shape, then a photograph taken at a later time t will be identical with that at $t = 0$, except that the wave profile has moved a distance ct in the positive direction of the x axis. If we took a new origin at the point $x = ct$, and let distances measured from this origin be called X, so that $x = X + ct$, then the equation of the wave profile referred to this new origin would be

$$\phi = f(X).$$

Referred to the original fixed origin, this means that

$$\phi = f(x - ct). \tag{1}$$

This equation is the most general expression of a wave moving with constant speed c and without change of shape, along the positive direction of x. If the wave is travelling in the negative direction its form is given by (1) with the sign of c changed, so that then

$$\phi = f(x + ct). \tag{2}$$

§3   Harmonic waves

The simplest example of a wave of this kind is the **harmonic wave**, in which the wave profile is a sine or cosine curve. Thus if the wave profile at $t = 0$ is

$$(\phi)_{t=0} = a \cos mx,$$

and the wave moves to the right with the constant speed c then, at time t, the displacement, or disturbance, is

$$\phi = a \cos m(x - ct). \tag{3}$$

The maximum modulus of the disturbance a is called the **amplitude**. This type of wave profile repeats itself at regular distances $2\pi/m$. The distance $2\pi/m$ is known as the **wavelength** λ of this periodic wave profile. Equation (3) could therefore be written

$$\phi = a \cos \frac{2\pi}{\lambda}(x - ct). \tag{4}$$

The time taken for one complete wave to pass any point is called the **period** τ of the wave. It follows from (4) that the argument $\dfrac{2\pi}{\lambda}(x - ct)$

must pass through a complete cycle of values as t is increased by τ. Thus from the periodicity of the cosine function we conclude that

$$\frac{2\pi c\tau}{\lambda} = 2\pi,$$

and so

$$\tau = \lambda/c. \tag{5}$$

The **frequency** n of such a periodic wave is the number of waves passing a fixed observer in unit time. Clearly

$$n = 1/\tau, \tag{6}$$

so that

$$c = n\lambda, \tag{7}$$

and equation (4) may thus be written in either of the equivalent forms,

$$\phi = a \cos 2\pi\left(\frac{x}{\lambda} - \frac{t}{\tau}\right), \tag{8}$$

or

$$\phi = a \cos 2\pi\left(\frac{x}{\lambda} - nt\right). \tag{9}$$

Sometimes it is useful to introduce the **wave number** k, which is the number of waves in a unit distance. Then

$$k = 1/\lambda, \tag{10}$$

and we may write equation (9)

$$\phi = a \cos 2\pi(kx - nt). \tag{11}$$

If we compare two similar waves

$$\phi_1 = a \cos 2\pi(kx - nt),$$

$$\phi_2 = a \cos \{2\pi(kx - nt) + \varepsilon\},$$

we see that ϕ_2 is the same as ϕ_1 except that it is displaced a distance $\varepsilon/2\pi k$ or, equivalently, $\varepsilon\lambda/2\pi$. The quantity ε is called the **phase** of ϕ_2 relative to ϕ_1. If $\varepsilon = 2\pi, 4\pi, \ldots$ then the displacement is exactly one, two, $\ldots$ wavelengths, and we say that the waves are *in phase*; if $\varepsilon = \pi, 3\pi, \ldots$ then the two waves are said to be exactly *out of phase*.

Even if a wave is not an harmonic wave, but the wave profile consists of a regularly repeating pattern, the definitions of wavelength, period, frequency and wave number still apply, and equations (5), (6), (7) and (10) are still valid.

§4   Plane waves

It is possible to generalise equation (1) to deal with the case of plane waves in three dimensions. A **plane wave** is one in which the disturbance is constant over all points of a plane drawn perpendicular to the direction of propagation. Such a plane is often called a **wavefront**, and this wavefront moves perpendicular to itself with the velocity of propagation c.

Suppose that the unit normal along the direction of propagation is $\boldsymbol{\nu}$ and that it has direction cosines (l, m, n). Then if $\mathbf{r}$, with components (x, y, z), is the position vector of a general point P on the plane wavefront at time t, we see from Fig. 1 that the equation of this wavefront is

$$\boldsymbol{\nu} \cdot \mathbf{r} = lx + my + nz = p,$$

where p is the perpendicular distance from the origin O measured along the vector $\boldsymbol{\nu}$ to the point Q at which this line meets the wavefront.

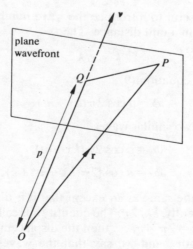

Fig. 1

As the plane wavefront moves with constant speed c along $\boldsymbol{\nu}$, it follows that if Q is distant α from O at time $t = 0$, so that $p = \alpha + ct$, the above result may be written

$$lx + my + nz - ct = \alpha \,(\text{cont}).\tag{12}$$

If, at any moment t, ϕ is to be constant for all x, y, z satisfying (12), then it is clear that

$$\phi = f(lx + my + nz - ct),$$

or

$$\phi = f(\boldsymbol{\nu}\,.\,\mathbf{r} - ct),\tag{13}$$

is a function which fulfils all these requirements and therefore represents a plane wave travelling with speed c in the direction $l:m:n$ without change of form. When, later, nonlinear waves are discussed, we shall see that the name wavefront is also used to characterise a somewhat different, though related, property of a wave.

§5   The wave equation

The expression (13) is a particular solution of the wave equation referred to on page 1. Since l, m, n are direction cosines, $l^2 + m^2 + n^2 = 1$, and it is easily verified that when f is twice differentiable the function ϕ satisfies the differential equation

$$\nabla^2\phi \equiv \frac{\partial^2\phi}{\partial x^2} + \frac{\partial^2\phi}{\partial y^2} + \frac{\partial^2\phi}{\partial z^2} = \frac{1}{c^2}\frac{\partial^2\phi}{\partial t^2},\tag{14}$$

which bears a close resemblance to Laplace's equation. This is the equation which is called the **wave equation**. It is one of the most important linear partial differential equations in the whole of mathematics, since it represents all types of linear wave motion in which the velocity is constant. The expressions in (1), (2), (8), (9), (11) and (13) are all particular solutions of this equation. We shall find, as we investigate different types of wave motion in subsequent chapters, that equation (14) invariably appears, and it will be our task to select the solution that is appropriate to our particular problem. There are certain types of solution that occur often, and we shall discuss some of them in the rest of this chapter, but before doing so, there is one important property of the fundamental equation to which reference has already been made that must be explained further.

§6   Principle of superposition

The wave equation is *linear*. That is to say, ϕ and its partial derivatives never occur in any form other than that of the first degree. Consequently, if ϕ_1 and ϕ_2 are any two solutions of (14), $a_1\phi_1 + a_2\phi_2$ is also a solution, a_1 and a_2 being two arbitrary constants. This is an illustration of the **principle of superposition**, which states that, when all the relevant equations are linear we may superpose any number of individual solutions to form new functions which are themselves also solutions. We shall often have occasion to do this and it is, indeed, the fundamental principle on which methods for seeking solutions to the wave equation are based.

A particular instance of this superposition, which is important in many problems, comes by adding together two harmonic waves going in opposite directions with the same amplitude and speed. Thus, with two equal amplitude waves similar to (11) moving in opposite directions, we obtain

$$\phi = a \cos 2\pi(kx - nt) + a \cos 2\pi(kx + nt)$$
$$= 2a \cos 2\pi kx \cos 2\pi nt. \tag{15}$$

This is known as a **stationary wave**, to distinguish it from the earlier **progressive waves**. It owes its name to the fact that the wave profile does not move forward. In fact, ϕ always vanishes at the points for which $\cos 2\pi kx = 0$; that is for

$$x = \pm\frac{1}{4k}, \pm\frac{3}{4k}, \pm\frac{5}{4k}, \ldots.$$

These points are called the **nodes** of the wave and the intermediate points, where the amplitude $2a \cos 2\pi kx$ of ϕ is greatest, are called **antinodes**. The distance between successive nodes, or successive antinodes, is $1/2k$, which, by (10), is half a wavelength.

Using equal amplitude harmonic wave functions similar to (13), we find corresponding stationary waves in three dimensions, given by

$$\phi = a \cos\{2\pi(\boldsymbol{\nu}\cdot\mathbf{r} - ct)/\lambda\} + a \cos\{2\pi(\boldsymbol{\nu}\cdot\mathbf{r} + ct)/\lambda\}$$
$$= 2a \cos(2\pi\boldsymbol{\nu}\cdot\mathbf{r}/\lambda) \cos(2\pi ct/\lambda). \tag{16}$$

In this case ϕ always vanishes on the planes

$$\boldsymbol{\nu}\cdot\mathbf{r} = lx + my + nz = \pm(2m + 1)\lambda/4,$$

where $m = 0, 1, \ldots$, and these are known as **nodal planes**.

§7   Special types of solution

We shall now obtain some special types of solution of the wave equation which we shall be able to apply to specific problems in later chapters. We may divide our solutions into two main types, representing stationary and progressive waves.

We have already dealt with progressive waves in one dimension. The equation to be solved is

$$\frac{\partial^2 \phi}{\partial x^2} = \frac{1}{c^2}\frac{\partial^2 \phi}{\partial t^2}.$$

Its most general solution may be obtained by a method due to d'Alembert. We change to new variables $u = x - ct$, and $v = x + ct$. Then it is easily verified that under this transformation

$$\frac{\partial \phi}{\partial x} = \frac{\partial \phi}{\partial u} + \frac{\partial \phi}{\partial v}, \qquad \frac{\partial \phi}{\partial t} = -c\frac{\partial \phi}{\partial u} + c\frac{\partial \phi}{\partial v}$$

$$\frac{\partial^2 \phi}{\partial x^2} = \frac{\partial^2 \phi}{\partial u^2} + 2\frac{\partial^2 \phi}{\partial u\,\partial v} + \frac{\partial^2 \phi}{\partial v^2}, \qquad \frac{\partial^2 \phi}{\partial t^2} = c^2\frac{\partial^2 \phi}{\partial u^2} - 2c^2\frac{\partial^2 \phi}{\partial u\,\partial v} + c^2\frac{\partial^2 \phi}{\partial v^2}.$$

When these changes are made the wave equation becomes

$$\frac{\partial^2 \phi}{\partial u\,\partial v} = 0.$$

The most general solution of this is

$$\phi = f(u) + g(v),$$

f and g being arbitrary twice differentiable functions of their arguments. In the original variables this is

$$\phi = f(x - ct) + g(x + ct). \tag{17}$$

The harmonic waves of §2 are special cases of this, in which f and g are cosine functions. The waves f and g travel with speed c, in opposite directions, with f moving to the right and g to the left.

In two dimensions the wave equation is

$$\frac{\partial^2 \phi}{\partial x^2} + \frac{\partial^2 \phi}{\partial y^2} = \frac{1}{c^2}\frac{\partial^2 \phi}{\partial t^2}. \tag{18}$$

By an obvious extension of d'Alembert's method it follows that the most general solution involving only plane waves is

$$\phi = f(lx + my - ct) + g(lx + my + ct), \tag{19}$$

where, as before, f and g are arbitrary twice differentiable functions of their arguments and $l^2 + m^2 = 1$. Strictly speaking, these should be called line waves, since at any moment ϕ is constant along the lines $lx + my = $ const.

In three dimensions the partial differential equation is

$$\frac{\partial^2 \phi}{\partial x^2} + \frac{\partial^2 \phi}{\partial y^2} + \frac{\partial^2 \phi}{\partial z^2} = \frac{1}{c^2} \frac{\partial^2 \phi}{\partial t^2}, \tag{20}$$

and the most general solution involving only plane waves is

$$\phi = f(lx + my + nz - ct) + g(lx + my + nz + ct), \tag{21}$$

in which $l^2 + m^2 + n^2 = 1$ and f, g satisfy the same differentiability condition as before.

There are, however, other solutions of progressive type, not involving plane waves. For suppose that we transform (20) to spherical polar coordinates r, θ, ψ, then the three-dimensional wave equation becomes

$$\frac{\partial^2 \phi}{\partial r^2} + \frac{2}{r} \frac{\partial \phi}{\partial r} + \frac{1}{r^2 \sin \theta} \frac{\partial}{\partial \theta}\left(\sin \theta \frac{\partial \phi}{\partial \theta}\right) + \frac{1}{r^2 \sin^2 \theta} \frac{\partial^2 \phi}{\partial \psi^2} = \frac{1}{c^2} \frac{\partial^2 \phi}{\partial t^2}. \tag{22}$$

If we are interested in solutions possessing spherical symmetry (i.e. independent of θ and ψ) we shall have to solve the simpler equation

$$\frac{\partial^2 \phi}{\partial r^2} + \frac{2}{r} \frac{\partial \phi}{\partial r} = \frac{1}{c^2} \frac{\partial^2 \phi}{\partial t^2}. \tag{23}$$

This may be written

$$\frac{\partial^2}{\partial r^2}(r\phi) = \frac{1}{c^2} \frac{\partial^2}{\partial t^2}(r\phi),$$

showing (cf. equation (17)) that it has solutions

$$r\phi = f(r - ct) + g(r + ct),$$

where f and g are again arbitrary twice differentiable functions of their arguments. We see, therefore, that there are progressive type solutions

$$\phi = \frac{1}{r}f(r - ct) + \frac{1}{r}g(r + ct). \tag{24}$$

Let us now turn to solutions of stationary type. These may all be obtained by the method known as the **separation of variables**. In one dimension we have to solve

$$\frac{\partial^2 \phi}{\partial x^2} = \frac{1}{c^2} \frac{\partial^2 \phi}{\partial t^2}$$

Let us try to find a solution of the form

$$\phi = X(x)T(t),$$

X and T being functions solely of x and t, respectively, whose form is still to be discovered. Substituting this form of ϕ in the differential equation and dividing both sides by the product $X(x)T(t)$ we obtain an equation involving only ordinary derivatives

$$\frac{1}{X} \frac{d^2 X}{dx^2} = \frac{1}{c^2 T} \frac{d^2 T}{dt^2}. \tag{25}$$

The left-hand side is independent of t, being only a function of x, and the right-hand side is independent of x, being only a function of t. Since the two sides are identically equal, this implies that each is independent both of x and t, and must therefore be constant. Putting this constant equal to $-p^2$, we arrive at the two equations

$$X'' + p^2 X = 0, \; T'' + c^2 p^2 T = 0. \tag{26}$$

Employing a concise but obvious notation, we see that these equations give, apart from arbitrary constants,

$$X = \frac{\cos}{\sin} px, \quad T = \frac{\cos}{\sin} cpt, \tag{27}$$

which are to be read X equals cos or sin px, etc. A typical solution therefore is $a \cos px \cos cpt$, in which p is arbitrary. In this expression we could replace either or both of the cosines by sines, and by the principle of superposition the complete solution is the sum of any number of terms of this kind with different values of p.

The constant $-p^2$ which we introduced, is known as the **separation constant**. We were able to introduce it in (25) because the variables x and t had been completely separated from each other and were in fact on opposite sides of the equation. There was no reason why the separation constant should have had a negative value of $-p^2$ except that this enabled us to obtain harmonic solutions (27). If we had set each side of (25) equal to $+p^2$, the solutions would have been

$$X = \exp{(\pm px)}, \qquad T = \exp{(\pm cpt)}, \tag{28}$$

This method of separation of variables can be extended to any number of dimensions. Thus in two dimensions a typical solution of (18) is

$$\phi = \frac{\cos}{\sin}\, px \, \frac{\cos}{\sin}\, qy \, \frac{\cos}{\sin}\, rct, \tag{29}$$

in which $p^2 + q^2 = r^2$, p and q being allowed arbitrary values. An alternative version of (29), in which one of the functions is exponential, is

$$\phi = \frac{\cos}{\sin}\, px \, \exp{(\pm qy)}\, \frac{\cos}{\sin}\, rct, \tag{30}$$

in which $p^2 - q^2 = r^2$.

It is easy to see that there is a variety of forms similar to (30) in which one or more of the functions is altered from an harmonic to a hyperbolic or exponential term.

In three dimensions we have solutions of the same type, two typical examples being

$$\phi = \frac{\cos}{\sin}\, px \, \frac{\cos}{\sin}\, qy \, \frac{\cos}{\sin}\, rz \, \frac{\cos}{\sin}\, sct, \qquad p^2 + q^2 + r^2 = s^2 \tag{31}$$

and

$$\phi = \frac{\cosh}{\sinh}\, px \, \exp{(\pm qy)}\, \frac{\cos}{\sin}\, rz \, \frac{\cos}{\sin}\, sct, \qquad -p^2 - q^2 + r^2 = s^2. \tag{32}$$

There are two other examples of a solution in three dimensions that we shall discuss. In the first case we put $x = r\cos\theta$, $y = r\sin\theta$, and we use r, θ and z as cylindrical coordinates. The wave equation then becomes

$$\frac{\partial^2 \phi}{\partial r^2} + \frac{1}{r}\frac{\partial \phi}{\partial r} + \frac{1}{r^2}\frac{\partial^2 \phi}{\partial \theta^2} + \frac{\partial^2 \phi}{\partial z^2} = \frac{1}{c^2}\frac{\partial^2 \phi}{\partial t^2}.$$

A solution can be found of the form

$$\phi = R(r)\Theta(\theta)Z(z)T(t), \tag{33}$$

where, by the method of separation of variables, R, Θ, Z, T are functions only of r, θ, z and t, respectively, and satisfy the equations

$$\frac{d^2R}{dr^2} + \frac{1}{r}\frac{dR}{dr} - \frac{m^2}{r^2}R + n^2R = 0, \qquad \frac{d^2\Theta}{d\theta^2} = -m^2\Theta,$$

$$\frac{d^2Z}{dz^2} = -q^2Z, \qquad \frac{d^2T}{dt^2} = -c^2p^2T, n^2 = p^2 - q^2. \qquad (34)$$

The only difficult equation is the first, and this is just **Bessel's equation** of order m, with solutions $J_m(nr)$ and $Y_m(nr)$. J_m is finite and Y_m is infinite when $r = 0$, so if a solution is required which is finite at the origin we shall require only the J_m solutions. The final form of ϕ is therefore

$$\phi = \frac{J_m}{Y_m}(nr)\frac{\cos}{\sin}m\theta\frac{\cos}{\sin}qz\frac{\cos}{\sin}cpt. \qquad (35)$$

If ϕ is to be single valued, m must be an integer; but n, q and p may be arbitrary provided that $n^2 = p^2 - q^2$. Hyperbolic modifications of (35) are possible, similar in all respects to (31) and (32).

Our final solution is one in spherical polar coordinates r, θ, ψ. The wave equation (22) has a solution $R(r)\Theta(\theta)\Psi(\psi)T(t)$, where

$$\frac{d^2T}{dt^2} = -c^2p^2T, \qquad \frac{d^2\Psi}{d\psi^2} = -m^2\Psi,$$

$$\frac{1}{\sin\theta}\frac{d}{d\theta}\left(\sin\theta\frac{d\Theta}{d\theta}\right) + \left\{n(n+1) - \frac{m^2}{\sin^2\theta}\right\}\Theta = 0,$$

$$\frac{d^2R}{dr^2} + \frac{2}{r}\frac{dR}{dr} + \left\{p^2 - \frac{n(n+1)}{r^2}\right\}R = 0.$$

Here m, n and p are arbitrary constants, but if $\Psi(\psi)$ is to be single valued, m must be integral. The first two of these equations present no difficulties. The θ-equation is the generalised **Legendre's equation** with solution

$$\Theta(\theta) = P_n^m(\cos\theta),$$

and if Θ is to be finite everywhere, n must be a positive integer. When $m = 0$ and n is integral, $P_n^m(\cos\theta)$ reduces to a polynomial in $\cos\theta$ of degree n, known as the *Legendre polynomial $P_n(\cos\theta)$*. For other integral values of m, $P_n^m(\cos\theta)$ is defined by the equation

$$P_n^m(\cos\theta) = \sin^m\theta\frac{d^m}{d(\cos\theta)^m}\{P_n(\cos\theta)\}.$$

A few values of $P_n(\cos\theta)$ and $P_n^m(\cos\theta)$ are given below, for small integral values of n and m. When $m > n$, $P_n^m(\cos\theta)$ vanishes identically.

$$P_0(\cos\theta) = 1$$

$$P_1(\cos\theta) = \cos\theta$$

$$P_2(\cos\theta) = \tfrac{1}{2}(3\cos^2\theta - 1)$$

$$P_3(\cos\theta) = \tfrac{1}{2}(5\cos^3\theta - 3\cos\theta)$$

$$P_4(\cos\theta) = \tfrac{1}{6}(35\cos^4\theta - 30\cos^2\theta + 3)$$

$$P_1^1(\cos\theta) = \sin\theta$$

$$P_2^1(\cos\theta) = 3\sin\theta\cos\theta$$

$$P_3^1(\cos\theta) = \tfrac{3}{2}\sin\theta\,(5\cos^2\theta - 1)$$

$$P_2^2(\cos\theta) = 3\sin^2\theta.$$

To solve the R-equation put $R(r) = r^{-1/2}S(r)$, and we find that the equation for $S(r)$ is just Bessel's equation

$$\frac{d^2 S}{dr^2} + \frac{1}{r}\frac{dS}{dr} + \left\{p^2 - \frac{(n+\frac{1}{2})^2}{r^2}\right\}S = 0,$$

so that

$$S(r) = J_{n+1/2}(pr) \quad \text{or} \quad Y_{n+1/2}(pr).$$

Collecting the various terms, the complete solution, apart from hyperbolic modifications, is seen to be

$$\phi = r^{-1/2}\frac{J_{n+1/2}}{Y_{n+1/2}}(pr)P_n^m(\cos\theta)\frac{\cos}{\sin}m\psi\frac{\cos}{\sin}cpt. \tag{36}$$

If ϕ has axial symmetry, we must only take functions with $m = 0$, and if it has spherical symmetry, terms with $m = n = 0$. Now $J_{1/2}(z) = \sqrt{(2/\pi z)}\sin z$, and also $Y_{1/2}(z) = -\sqrt{(2/\pi z)}\cos z$, so that for $m = n = 0$ this becomes

$$\phi = r^{-1}\frac{\cos}{\sin}pr\frac{\cos}{\sin}cpt. \tag{37}$$

A solution finite at the origin is obtained by omitting the cos pr term.

§8   List of solutions

We shall now gather together for future reference the solutions obtained in the preceding pages.

Progressive waves

1 dimension

$$\frac{\partial^2 \phi}{\partial x^2} = \frac{1}{c^2}\frac{\partial^2 \phi}{\partial t^2}, \qquad \phi = f(x-ct) + g(x+ct) \tag{17}$$

2 dimensions

$$\frac{\partial^2 \phi}{\partial x^2} + \frac{\partial^2 \phi}{\partial y^2} = \frac{1}{c^2}\frac{\partial^2 \phi}{\partial t^2},$$

$$\phi = f(lx + my - ct) + g(lx + my + ct), l^2 + m^2 = 1 \tag{19}$$

3 dimensions

$$\frac{\partial^2 \phi}{\partial x^2} + \frac{\partial^2 \phi}{\partial y^2} + \frac{\partial^2 \phi}{\partial z^2} = \frac{1}{c^2}\frac{\partial^2 \phi}{\partial t^2},$$

$$\phi = f(lx + my + nz - ct) + g(lx + my + nz + ct), l^2 + m^2 + n^2 = 1 \tag{21}$$

3 dimensions, spherical symmetry

$$\frac{\partial^2 \phi}{\partial r^2} + \frac{2}{r}\frac{\partial^2 \phi}{\partial t^2} = \frac{1}{c^2}\frac{\partial^2 \phi}{\partial t^2}, \qquad \phi = \frac{1}{r}f(r-ct) + \frac{1}{r}g(r+ct). \tag{24}$$

Stationary waves

1 dimension

$$\frac{\partial^2 \phi}{\partial x^2} = \frac{1}{c^2}\frac{\partial^2 \phi}{\partial t^2},$$

$$\phi = \frac{\cos}{\sin}px\,\frac{\cos}{\sin}cpt \tag{27}$$

$$\phi = \exp(\pm px \pm cpt) \tag{28}$$

2 dimensions

$$\frac{\partial^2 \phi}{\partial x^2} + \frac{\partial^2 \phi}{\partial y^2} = \frac{1}{c^2}\frac{\partial^2 \phi}{\partial t^2},$$

$$\phi = \frac{\cos}{\sin}px\,\frac{\cos}{\sin}qy\,\frac{\cos}{\sin}rct, p^2 + q^2 = r^2 \tag{29}$$

$$\phi = \frac{\cos}{\sin}px\,\exp(\pm qy)\,\frac{\cos}{\sin}rct, p^2 - q^2 = r^2 \tag{30}$$

3 dimensions

$$\frac{\partial^2 \phi}{\partial x^2}+\frac{\partial^2 \phi}{\partial y^2}+\frac{\partial^2 \phi}{\partial z^2}=\frac{1}{c^2}\frac{\partial^2 \phi}{\partial t^2},$$

$$\phi=\frac{\cos}{\sin}px\,\frac{\cos}{\sin}qy\,\frac{\cos}{\sin}rz\,\frac{\cos}{\sin}sct,$$

$$p^2+q^2+r^2=s^2 \qquad (31)$$

$$\phi=\frac{\cosh}{\sinh}px\,\exp\,(\pm qy)\,\frac{\cos}{\sin}rz\,\frac{\cos}{\sin}sct,$$

$$-p^2-q^2+r^2=s^2. \qquad (32)$$

Plane Polar Coordinates (r,θ)

$$\frac{\partial^2 \phi}{\partial r^2}+\frac{1}{r}\frac{\partial \phi}{\partial r}+\frac{1}{r^2}\frac{\partial^2 \phi}{\partial \theta^2}=\frac{1}{c^2}\frac{\partial^2 \phi}{\partial t^2}, \qquad \phi=\frac{J_m}{Y_m}(nr)\,\frac{\cos}{\sin}m\theta\,\frac{\cos}{\sin}cnt. \qquad (35a)$$

Cylindrical Polar Coordinates (r,θ,z)

$$\frac{\partial^2 \phi}{\partial r^2}+\frac{1}{r}\frac{\partial \phi}{\partial r}+\frac{1}{r^2}\frac{\partial^2 \phi}{\partial \theta^2}+\frac{\partial^2 \phi}{\partial z^2}=\frac{1}{c^2}\frac{\partial^2 \phi}{\partial t^2},$$

$$\phi=\frac{J_m}{Y_m}(nr)\,\frac{\cos}{\sin}m\theta\,\frac{\cos}{\sin}qz\,\frac{\cos}{\sin}cpt, \qquad n^2=p^2-q^2. \qquad (35b)$$

Spherical Polar Coordinates (r,θ,ψ)

$$\frac{\partial^2 \phi}{\partial r^2}+\frac{2}{r}\frac{\partial \phi}{\partial r}+\frac{1}{r^2\sin\theta}\frac{\partial}{\partial \theta}\left(\sin\theta\frac{\partial \phi}{\partial \theta}\right)+\frac{1}{r^2\sin^2\theta}\frac{\partial^2 \phi}{\partial \psi^2}=\frac{1}{c^2}\frac{\partial^2 \phi}{\partial t^2},$$

$$\phi=r^{-1/2}\frac{J_{n+1/2}}{Y_{n+1/2}}(pr)P_n^m(\cos\theta)\,\frac{\cos}{\sin}m\psi\,\frac{\cos}{\sin}cpt. \qquad (36)$$

Spherical symmetry

$$\frac{\partial^2 \phi}{\partial r^2}+\frac{2}{r}\frac{\partial \phi}{\partial r}=\frac{1}{c^2}\frac{\partial^2 \phi}{\partial t^2}, \qquad \phi=r^{-1}\frac{\cos}{\sin}pr\,\frac{\cos}{\sin}cpt. \qquad (37)$$

It should be noted that there are exponential modifications of all the above equations.

In solving problems, we shall more often require progressive type solutions in cases where the variables x, y, z are allowed an infinite range of values, and stationary type solutions when their allowed range is finite.

§9 Equation of telegraphy

There is an important modification of the wave equation which arises when friction, or some other dissipative force, produces a damping. The damping effect is usually allowed for (see e.g. Chapter 2 and elsewhere) by the inclusion of a term of the form $p\dfrac{\partial\phi}{\partial t}$, which will arise when the damping force is proportional to the velocity of the vibrations. The revised form of the fundamental equation of wave motion, known as the **equation of telegraphy**, is

$$\nabla^2\phi = \frac{1}{c^2}\left\{\frac{\partial^2\phi}{\partial t^2} + p\frac{\partial\phi}{\partial t}\right\}. \tag{38}$$

If we omit the term $\dfrac{\partial^2\phi}{\partial t^2}$ this equation is the same as that occurring in the flow of heat, though the mathematical properties of the resulting equation are then quite different and do not characterise wave propagation. If in (38) we set $\phi = u\exp(-pt/2)$, we obtain an equation for u of the form

$$\nabla^2 u = \frac{1}{c^2}\left\{\frac{\partial^2 u}{\partial t^2} - \frac{1}{4}p^2 u\right\}. \tag{39}$$

Very often p is so small that we may neglect p^2, and then (39) is in the standard form which we have discussed in **§8**, and the solutions given there will apply. In such a case the presence of the dissipative term is shown by a decay factor $\exp(-pt/2)$. If this is written in the form $\exp(-t/t_0)$, then $t_0(=2/p)$ is called the **modulus of decay**. When the term in p^2 may not be neglected, we have to solve (38), and the method of separation of variables usually enables a satisfactory solution to be obtained without much difficulty.

There is an alternative solution to the equation of telegraphy that is sometimes useful. Taking the case of one dimension, and supposing that p is so small that p^2 may be neglected, we have shown that the solution of (38) may be written in the form

$$\phi = \exp(-pt/2)f(x-ct), \tag{40}$$

where f is any twice differentiable function of its argument. Since f is otherwise arbitrary, we can put

$$f(x-ct) = \exp\{-p(x-ct)/2c\}g(x-ct),$$

where g is now an arbitrary function with the same differentiability properties as f. Substituting this in (40) we get

$$\phi = \exp{(-px/2c)}g(x-ct). \tag{41}$$

This expression resembles (40) except that the exponential factor now varies with x instead of with t.

§10   Exponential form of harmonic waves

Most of the waves with which we shall be concerned in later chapters will be harmonic. This is partly because, as we have seen in §8, harmonic functions arise very naturally when we try to solve the wave equation; it is also due to the fact that by means of a Fourier analysis, any function may be split into harmonic components, and hence by the principle of superposition, any wave máy be regarded as the resultant of a set of harmonic waves.

When dealing with progressive waves of harmonic type there is one simplification that is often useful and which is especially important in the electromagnetic theory of light waves. We have seen in (11) that a progressive harmonic wave in one dimension can be represented by $\phi = a \cos 2\pi(kx - nt)$. If we allow for a phase ε, it will be written $\phi = a \cos\{2\pi(kx - nt) + \varepsilon\}$. Now this latter function may be regarded as the real part of the complex quantity

$$a \exp\{\pm i[2\pi(kx - nt) + \varepsilon]\}.$$

It is most convenient for much of our subsequent work if we choose the minus sign and also absorb the phase ε and the amplitude a into one complex number A. We shall then write

$$\phi = A \exp\{2\pi i(nt - kx)\}, \qquad A = a \exp{(-i\varepsilon)}. \tag{42}$$

This complex quantity is itself a solution of the wave equation, as can easily be seen by substitution, and consequently both its real and imaginary parts are also solutions. Since all our equations in ϕ are linear, it is possible to use (42) itself as a solution of the wave equation, instead of its real part. In any equation in which ϕ appears to the first degree, we can, if we wish, use the function (42) and assume that we always refer to the real part, or we can just use (42) as it stands, without reference to its real or imaginary parts. In such a case the apparent amplitude A is usually complex, and since $A = a \exp{(-i\varepsilon)}$, we can say that $|A|$ is the true amplitude, and $-\arg A$ is the true phase. The speed, of course, as given by (7) and (10), is n/k.

We can extend this representation of ϕ to cover waves travelling in the opposite direction by using in such a case

$$\phi = A \exp\{2\pi i(nt + kx)\}. \tag{43}$$

There is obviously no reason why we should not extend this to two or three dimensions. For instance, in three dimensions

$$\phi = A \exp\{2\pi i(nt - \boldsymbol{\nu}.\mathbf{r})\} \tag{44}$$

would represent a harmonic wave with amplitude A moving with speed n in the positive direction of the unit vector $\boldsymbol{\nu}$.

Before concluding this section let us consider a generalisation of the one-dimensional form of the equation of telegraphy (38) in terms of the plane wave representation (42). To be precise, we shall consider the **generalised equation of telegraphy**

$$\frac{\partial^2 \phi}{\partial x^2} = \frac{1}{c^2}\left\{\frac{\partial^2 \phi}{\partial t^2} + p\frac{\partial \phi}{\partial t} + q\phi\right\} \tag{45}$$

where q, like p, is a constant. Then, clearly, a plane wave corresponding to an arbitrary choice of n and k in (42) cannot satisfy (45), so to find the relationship that they must satisfy it is necessary to substitute (42) into (45). When this is done the compatibility relation is obtained in the form

$$4\pi^2 n^2 - i2\pi pn - (4\pi^2 c^2 k^2 + q) = 0, \tag{46}$$

so that the frequency n is seen to become *complex* for real k, with

$$n = \frac{ip}{4\pi} \pm \frac{1}{4\pi}\{16\pi^2 c^2 k^2 + (4q - p^2)\}^{1/2}. \tag{47}$$

The plane wave (42) then takes the form

$$\phi = A \exp(-pt/2)\exp 2\pi i\left\{\pm\frac{t}{4\pi}[16\pi^2 c^2 k^2 + (4q - p^2)]^{1/2} - kx\right\}. \tag{48}$$

It now follows that if $p > 0$ the wave will attenuate with time, but since from (47) the frequency depends on the wave length λ through the wave number k, it also follows that the speed of propagation of the wave is frequency dependent. Thus the effect of equation (45) on the propagation of two initially coincident plane waves with different frequencies will be to separate them as time increases. By analogy with optics, this frequency dependence of the propagation speed is called

dispersion, and the compatibility relation (46) itself connecting n and k is known as the **dispersion relation** for differential equation (45). If $p < 0$ the solution ϕ in (48) will be **unstable** since it will grow without bound.

Dispersion in a partial differential equation representing wave motion leads directly to change of shape of the wave as it propagates. This can easily be seen by considering an initial wave form to be resolved into its Fourier components, when their different propagation speeds, corresponding to different frequencies, lead to a changed wave form when they are again superposed at some subsequent time.

In general equations of wave motion exhibit dispersion, and only in exceptional cases, as with the wave equation (14), will distortionless propagation be possible. A special exception is provided by the generalised equation of telegraphy (45) in the case that $4q = p^2$, for then (48) becomes

$$\phi = A \exp\left(-pt/2\right) \exp 2\pi ki(\pm ct - x). \tag{49}$$

This result shows that irrespective of k, all harmonic plane waves will propagate at the same speed c to the right or left without distortion, though they will all be equally attenuated by the factor $\exp\left(-pt/2\right)$. For this reason wave solutions to the telegraph equation in which $4q = p^2$ are called **relatively undistorted waves**. This condition is of considerable importance in telephone line construction where, if it is satisfied, the signal will be attenuated as it propagates but not distorted.

§11   D'Alembert's formula

Let us now consider the wave equation

$$\frac{\partial^2 \phi}{\partial t^2} - c^2 \frac{\partial^2 \phi}{\partial x^2} = 0 \tag{50}$$

together with the initial conditions

$$\phi(x, 0) = h(x) \quad \text{and} \quad \frac{\partial \phi}{\partial t}(x, 0) = k(x), \tag{51}$$

where h and k are arbitrarily assigned differentiable functions for $-\infty < x < \infty$. From the form of the solution that will subsequently be given in equation (57) it will appear that h must be at least twice differentiable and k at least once differentiable. The conditions (51) are called **initial conditions** because they involve the specification of the behaviour of the solution $\phi(x, t)$ at some specific moment in time, here taken to be $t = 0$. In the theory of partial differential equations this initial data is known as the **Cauchy data** for the wave equation.

Our purpose in this section will be to determine how the solution evolves away from these initial conditions as time increases. Since the initial or Cauchy data (51) is assumed to be given for all x, the problem represented by (50) and (51) is called a **pure initial value problem** for the wave equation for the *unbounded region* $-\infty < x < \infty$.

It is known from (17) that the general solution to (50), without considering the conditions (51), is

$$\phi = f(x - ct) + g(x + ct), \tag{52}$$

where f and g are arbitrary twice differentiable functions. So, setting $t = 0$ in (52), we have at once from the first condition in (51) that

$$f(x) + g(x) = h(x). \tag{53}$$

It also follows by differentiation of (52) partially with respect to t, followed by setting $t = 0$ and using the second condition in (51), that

$$-cf'(x) + cg'(x) = k(x). \tag{54}$$

Integration of this last result from an arbitrary point a to x then yields

$$-f(x) + g(x) = \frac{1}{c} \int_a^x k(s) \, ds + g(a) - f(a).$$

Combination with (53) now gives

$$f(x) = \tfrac{1}{2} h(x) - \frac{1}{2c} \int_a^x k(s) \, ds - \tfrac{1}{2}(g(a) - f(a)), \tag{55}$$

and

$$g(x) = \tfrac{1}{2} h(x) + \frac{1}{2c} \int_a^x k(s) \, ds + \tfrac{1}{2}(g(a) - f(a)). \tag{56}$$

Replacing x by $x - ct$ in (55) and by $x + ct$ in (56) and adding gives, by virtue of (52),

$$\phi = \tfrac{1}{2} \left\{ h(x - ct) + h(x + ct) - \frac{1}{c} \int_a^{x-ct} k(s) \, ds + \frac{1}{c} \int_a^{x+ct} k(s) \, ds \right\}.$$

Absorbing the minus sign in the third term on the right-hand side by reversing the limits of integration and then combining the last two terms finally yields

$$\phi(x, t) = \frac{h(x - ct) + h(x + ct)}{2} + \frac{1}{2c} \int_{x-ct}^{x+ct} k(s) \, ds, \qquad (57)$$

which is the solution to our problem. This is called **d'Alembert's formula** and it is an important result since it provides valuable information about the nature of the solution ϕ and the way it depends on the Cauchy data (51).

The first conclusion that may be drawn from (57) is that the Cauchy data given in (51) is sufficient to specify a *unique* solution to the wave equation. To see this let us suppose, if possible, that two different solutions ϕ and ψ to (51) exist that both satisfy the same Cauchy data (51). Then, because of the linearity of the wave equation, the function $w = \phi - \psi$ will also be a solution of (50), and it will have for its Cauchy data on the initial line $t = 0$ the homogeneous conditions

$$w(x, 0) = 0 \quad \text{and} \quad \frac{\partial w}{\partial t}(x, 0) = 0,$$

corresponding to $h = k \equiv 0$. D'Alembert's formula then shows that $w(x, t) \equiv 0$ for all x, and $t \geqslant 0$, so that $\phi \equiv \psi$ and hence the solution is unique.

To study the way in which the solution depends on the Cauchy data let us now consider the solution at a general point (x_0, t_0) in the (x, t) plane with $t_0 > 0$. Then from (57) the solution at (x_0, t_0) is seen to be determined only by Cauchy data given on the finite interval $x_0 - ct_0 \leqslant x \leqslant x_0 + ct_0$ of the initial line. More precisely, it is only influenced by the functional values of h at the ends of this interval, and by the function k over the entire interval by virtue of the integral term in (57).

It is for this reason that the interval

$$x_0 - ct_0 \leqslant x \leqslant x_0 + ct_0 \quad \text{along} \quad t = 0$$

on the initial line is called the **domain of dependence** of the point (x_0, t_0). The triangular region with point (x_0, t_0) as vertex and the domain of dependence of this point as base is called the **domain of determinacy** that is associated with the domain of dependence. This is so called because the solution at each point of it is fully determined by the initial data that has been assigned to the domain of dependence.

An immediate consequence of (57) is that two sets of different initial data that coincide only in some interval D of the initial line $t = 0$ define two different solutions which are, however, identical in their common domain of determinacy associated with D.

If a domain of determinacy on which initial data is given is reduced to a single point P of the initial line, then the solution will only be determined at that one point. However, the data at P will *influence*, but not determine, the solution at points in the half plane $t > 0$ that lie in an open triangular region with P as vertex and with sides comprising the straight lines drawn through P in the direction of increasing time with gradients $dx/dt = \pm c$. It is on account of this that the region so defined is called the **range of influence** of the point P. These ideas are illustrated in Fig. 2(a),(b). A straight line through an arbitrary point

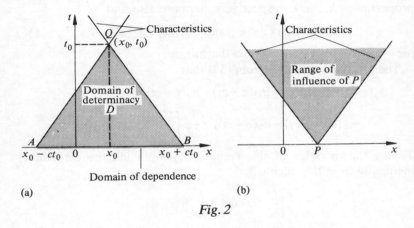

Fig. 2

$(\xi, 0)$ of the initial line that bounds one side of a domain of determinacy or a range of influence will have either the equation

$$x - ct = \xi \quad \text{or} \quad x + ct = \xi.$$

These lines belong to one of the two families of parallel straight lines $u = \text{const.}$ and $v = \text{const.}$, where

$$u = x - ct \quad \text{and} \quad v = x + ct,$$

which are collectively called the **families of characteristic curves** of the wave equation (50). Here we have expressly used the term *curve* rather than *line*, in relation to characteristics, since in more general situations where c is not constant the families of characteristics are indeed families of curves and not, as in this case, families of parallel straight lines.

To conclude this section let us consider two different initial value problems for the wave equation (50) that are determined by assigning the two different sets of Cauchy data

(i) $\phi_1(x, 0) = h_1(x)$ and $\dfrac{\partial \phi_1}{\partial t}(x, 0) = k_1(x)$

and

(ii) $\phi_2(x, 0) = h_2(x)$ and $\dfrac{\partial \phi_2}{\partial t}(x, 0) = k_2(x)$,

where h_1, h_2, k_1 and k_2 are functions with the same differentiability properties as h and k, respectively. Suppose also that

$$|h_1(x) - h_2(x)| < \varepsilon \quad \text{and} \quad |k_1(x) - k_2(x)| < \delta \tag{58}$$

for $-\infty < x < \infty$, with $\varepsilon > 0$, $\delta > 0$ arbitrary.

Then it follows directly from (57) that

$$\phi_1(x, t) - \phi_2(x, t) = \tfrac{1}{2}\{h_1(x - ct) - h_2(x - ct)\}$$
$$+ \tfrac{1}{2}\{h_1(x + ct) - h_2(x + ct)\} + \frac{1}{2c} \int_{x-ct}^{x+ct} (k_1(s) - k_2(s)) \, ds.$$

Taking the modulus of this result and employing the elementary inequality from the calculus

$$\left| \int_a^b f(s) \, ds \right| \leqslant \int_a^b |f(s)| \, ds$$

then gives

$$|\phi_1(x, t) - \phi_2(x, t)| \leqslant \tfrac{1}{2}|h_1(x - ct) - h_2(x - ct)|$$
$$+ \tfrac{1}{2}|h_1(x + ct) - h_2(x + ct)| + \frac{1}{2c} \int_{x-ct}^{x+ct} |k_1(s) - k_2(s)| \, ds.$$

So, using inequalities (58), this reduces to the result

$$|\phi_1(x, t) - \phi_2(x, t)| \leqslant \varepsilon + \delta t.$$

Consequently, for any given $t = \tau$, we conclude that by making h_2 and k_2 sufficiently close approximations to h_1 and k_1, so that ε, δ are suitably small, the solution ϕ_2 will approximate solution ϕ_1 arbitrarily closely for all x and $0 \leqslant t \leqslant \tau$.

Expressed differently, this fundamental result asserts that *the solution to the wave equation depends continuously on the Cauchy data*. This conclusion is in agreement with observations in the physical world, in the sense that if the Cauchy data is slightly altered, then the solution itself is only slightly changed. We shall have occasion to appeal to this result when we come to study the idealised motion of a plucked string.

§12   Inhomogeneous wave equation

The wave equation (50) is said to be **homogeneous**, in the sense that each term depends linearly on ϕ. It has associated with it an **inhomogeneous** equation

$$\frac{\partial^2 \phi}{\partial t^2} - c^2 \frac{\partial^2 \phi}{\partial x^2} = f(x, t) \tag{59}$$

in which, as we shall see in subsequent sections, the inhomogeneous term $f(x, t)$ which does not contain ϕ represents some externally acting force. We now set out to determine how the inhomogeneous term will modify d'Alembert's formula when (59) is subject to the initial conditions (51).

Our task may be simplified by observing at the outset that if in (59) we set $\phi = \phi_1 + \phi_2$, where ϕ_1 is a solution of the homogeneous wave equation (50) subject to the general initial data (51), then ϕ_2 will be a solution of

$$\frac{\partial^2 \phi_2}{\partial t^2} - c^2 \frac{\partial^2 \phi_2}{\partial x^2} = f(x, t) \tag{60}$$

subject to the homogeneous initial conditions

$$\phi_2(x, 0) = 0 \quad \text{and} \quad \frac{\partial \phi_2}{\partial t}(x, 0) = 0. \tag{61}$$

Henceforth, we need work only with the initial value problem represented by (60) and (61).

Our starting point will be to integrate (60) over the region D in Fig. 2a to obtain

$$\iint_D \left(\frac{\partial^2 \phi_2}{\partial t^2} - c^2 \frac{\partial^2 \phi_2}{\partial x^2} \right) dx \, dt = \iint_D f(x, t) \, dx \, dt.$$

Employing Green's theorem to replace the integral over D on the left-hand side by a line integral around the boundary ∂D of D then reduces this result to

$$-\oint_{\partial D} \frac{\partial \phi_2}{\partial t} \, dx + c^2 \frac{\partial \phi_2}{\partial x} \, dt = \iint_D f(x, t) \, dx \, dt. \qquad (62)$$

Now the boundary ∂D comprises the three directed straight line segments BQ, QA and AB in Fig. 2a, and along BQ $dx/dt = -c$, whilst along QB $dx/dt = c$. By virtue of these results (62) may be written

$$\oint_{BQ} c\left(\frac{\partial \phi_2}{\partial t} \, dt + \frac{\partial \phi_2}{\partial x} \, dx\right) - \oint_{QA} c\left(\frac{\partial \phi_2}{\partial t} \, dt + \frac{\partial \phi_2}{\partial x} \, dx\right)$$

$$-\oint_{AB} \left(\frac{\partial \phi_2}{\partial t} \, dx + c^2 \frac{\partial \phi_2}{\partial x} \, dt\right) = \iint_D f(x, t) \, dx \, dt. \qquad (63)$$

The bracketed terms in the first two integrals are simply the total differential $d\phi_2$, while in the third integrand the first term vanishes on AB because of the second initial condition in (61), and the second term vanishes because as AB is directed along the x-axis $dt/dx = 0$. Hence we arrive at the result

$$\oint_{BQ} c \, d\phi_2 - \oint_{QA} c \, d\phi_2 = \iint_D f(x, t) \, dx \, dt$$

so, recalling the directed nature of the segments BQ and QA of the boundary, we may write

$$c\phi_2(Q) - c\phi_2(B) + c\phi_2(Q) - c\phi_2(A) = \iint_D f(x, t) \, dx \, dt. \qquad (64)$$

The first of initial conditions (61) assert that $\phi_2(A) = \phi_2(B) = 0$, so that (64) becomes

$$\phi_2(Q) = \frac{1}{2c} \iint_D f(x, t) \, dx \, dt. \qquad (65)$$

Expressing the integral over D in terms of the geometry of triangle ABQ in Fig. 2a we finally arrive at the result

$$\phi_2(x_0, t_0) = \frac{1}{2c} \int_0^{t_0} \int_{x_0 - c(t_0 - t')}^{x_0 + c(t_0 - t')} f(x', t') \, dx' \, dt'. \qquad (66)$$

Dropping the suffix zero, and using the fact that $\phi = \phi_1 + \phi_2$ to combine d'Alembert's formula (57) and (66) then shows that the solution to the inhomogeneous wave equation (60) subject to the general initial conditions (51) is

$$\phi(x, t) = \frac{h(x - ct) + h(x + ct)}{2} + \frac{1}{2c} \int_{x-ct}^{x+ct} k(s)\, \mathrm{d}s$$

$$+ \frac{1}{2c} \int_0^t \int_{x-c(t-t')}^{x+c(t-t')} f(x', t')\, \mathrm{d}x'\, \mathrm{d}t'. \tag{67}$$

The form of the solution to the inhomogeneous problem still expresses the dependence of the solution on the Cauchy data given on the domain of dependence of the point (x, t) on the initial line $t = 0$. However, now the solution also depends on the behaviour of $f(x, t)$ at all points interior to the triangle with vertices at (x, t), $(x - ct, 0)$ and $(x + ct, 0)$ in the (x, t) plane. Consequently, for the inhomogeneous wave equation, it is appropriate to refer to this entire triangular region as the domain of dependence of the point (x, t).

§13   Boundary conditions and mixed problems

So far the region in which the solution has been obtained has been unbounded in space, for we have found $\phi(x, t)$ subject to the conditions $-\infty < x < \infty$ and $t \geqslant 0$. This situation is undesirably restrictive because problems often arise in which either the region involved is semi-infinite, so that $x \geqslant a$ and $t \geqslant 0$, or it is a bounded region in space so that, say, $a \leqslant x \leqslant b$ and $t \geqslant 0$. The line $x = a$ in the first case, and the lines $x = a$ and $x = b$ in the second case, then represent **spatial boundaries** for the regions concerned. On account of this, when the usual initial data is prescribed for $t = 0$ along that part of the x axis that lies within the region, such a problem subject to some conditions that are to be given on the boundaries is then called a **mixed initial and boundary value problem**. Let us now show that if ϕ is specified along these boundaries, the solution will be uniquely determined throughout the whole of the region that is involved.

To do this we first need a preliminary result that comes by considering the characteristic parallelogram in Fig. 3 whose sides comprise segments AB, BC, CD and DA of characteristic lines. Here the centre of the parallelogram is assumed to lie at the general point (x_0, t_0), while A is taken to be the point $(x_0 + \xi, t_0 + \eta)$ and C the point $(x_0 - \xi, t_0 - \eta)$

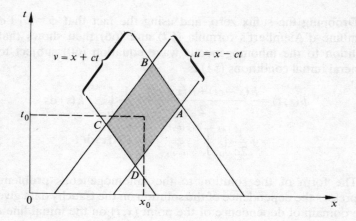

with ξ, η arbitrary. A simple calculation using the equations of the families of characteristics through A and C then shows that B is the point $(x_0 + c\eta, t_0 + \xi/c)$ and D the point $(x_0 - c\eta, t_0 - \xi/c)$. It is now merely a matter of elementary algebra to verify that as

$$\phi(x, t) = f(x - ct) + g(x + ct),$$

we must have

$$\phi(A) + \phi(C) = \phi(B) + \phi(D). \tag{68}$$

This useful result asserts that if ϕ is known at three corners of a characteristic parallelogram, then its value at the remaining point is determined uniquely by (68).

If, now, we consider Fig. 4, the implication of (68) on a mixed initial and boundary value problem becomes apparent. Suppose the semi-infinite region $x \geqslant 0$, $t \geqslant 0$ is involved and that the initial data (51) is specified on $t = 0$ for $x \geqslant 0$. Then from (57) the solution will be known only in the shaded region. Constructing the characteristic parallelogram $PQRS$ and employing (68) we see that if ϕ is known at points Q, R and S, then it must also be known at P. Consequently, if ϕ is specified along the spatial boundary $x = 0$, the solution may be found at all points of the unshaded region $x - ct > 0$ and $x > 0$, because from (57) it is known at all points R, S of the line $x - ct = 0$. As the solution is unique in the shaded region, and (68) is a linear relationship, the solution will also be specified uniquely throughout the semi-infinite region.

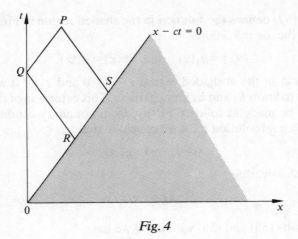

Fig. 4

A similar argument can be used to construct a unique solution in a bounded region $a \leq x \leq b, t \geq 0$ when initial data is given on $a \leq x \leq b, t = 0$ and ϕ is specified along $x = a$ and $x = b$.

§14   Extension of solutions by reflection

There is an alternative way of looking at the mixed initial and boundary value problem just discussed that does not employ result (68). In place of this it extends the problem, to a pure initial value problem requiring solution in the unbounded region $-\infty < x < \infty$, in such a way that the solution of the mixed initial and boundary value problem that is sought coincides with this solution in the region in question. We illustrate this approach by considering the mixed initial and boundary value problem for the homogeneous wave equation associated with the region $0 \leq x < \infty$ and $t \geq 0$ illustrated in Fig. 4. The initial conditions given on the positive half of the x-axis are taken to be

$$\phi(x, 0) = h_1(x) \quad \text{and} \quad \frac{\partial \phi}{\partial t}(x, 0) = k_1(x) \quad \text{for} \quad x \geq 0, \qquad (69)$$

and the boundary condition on $x = 0$ will be taken to be the homogeneous condition

$$\phi(0, t) = 0. \qquad (70)$$

Result (57) defines the solution in the shaded region in Fig. 4 when we make the identifications

$$h(x) = h_1(x) \quad \text{and} \quad k(x) = k_1(x).$$

To obtain ϕ in the unshaded region $x - ct > 0$ and $x > 0$ it would be necessary to know h_1 and k_1 for negative x. This extension of the initial data may be made as follows. Firstly, from boundary condition (70) and the general solution (52), we conclude that

$$0 = f(-ct) + g(ct)$$

or, equivalently, that

$$f(-x) = -g(x). \tag{71}$$

Using results (55) and (56) with $a = 0$ we have

$$f(x) = \tfrac{1}{2} h_1(x) - \frac{1}{2c} \int_0^x k_1(s) \, ds - \tfrac{1}{2}(g(0) - f(0)), \tag{72}$$

and

$$g(x) = \tfrac{1}{2} h_1(x) + \frac{1}{2c} \int_0^x k_1(s) \, ds + \tfrac{1}{2}(g(0) + f(0)). \tag{73}$$

Replacing x by $-x$ in (72) then gives

$$f(-x) = \tfrac{1}{2} h_1(-x) + \frac{1}{2c} \int_0^x k_1(-s) \, ds - \tfrac{1}{2}(g(0) - f(0)). \tag{74}$$

It now follows that (73) and (74) can only satisfy (71) if

$$h_1(-x) = -h_1(x) \quad \text{and} \quad k_1(-x) = -k_1(x), \tag{75}$$

thereby showing that if h_1 and k_1 are to be extended for negative x, they must both be extended as **odd functions**.

Thus the solution of the mixed initial and boundary value problem (69), (70) for the homogeneous wave equation coincides with the solution of the pure initial value problem

$$\phi(x, 0) = h(x) \quad \text{and} \quad \frac{\partial \phi}{\partial t}(x, 0) = k(x)$$

in the semi-infinite region $x \geq 0$, $t \geq 0$, when we set

$$h(x) = \begin{cases} h_1(x) \text{ for } x \geq 0 \\ -h_1(x) \text{ for } x < 0 \end{cases} \quad \text{and} \quad k(x) = \begin{cases} k_1(x) \text{ for } x \geq 0 \\ -k_1(x) \text{ for } x < 0. \end{cases}$$

This form of solution of the problem in terms of an associated pure initial value problem can be interpreted in terms of **reflection**. To do this it is first necessary to observe that in $x > 0, t \geq 0$ the solution comprises waves moving to the right and left. Our extension of the initial data for $x < 0$ then amounts to regarding the line $x = 0$ as a *reflecting barrier* with the property that when a wave moving to the left reaches it, a reflection takes place together with a change of sign.

The specification of ϕ on a boundary is called a **fixed boundary condition** to distinguish it from the so called **free boundary condition** in which $\partial \phi / \partial x$ is specified on a boundary. A unique solution to a free boundary problem may also be constructed from an associated pure initial value problem by using d'Alembert's formula and initial conditions deduced from equations (52), (72) and (73). The details of this are left as an exercise for the reader (see Example 17).

§15   A solved example

It is appropriate at this point that we should consider an example which illustrates an alternative method of approach to a mixed initial and boundary value problem. We choose to discuss one involving a finite region in the plane. Let us find a solution of

$$\frac{\partial^2 \phi}{\partial x^2} + \frac{\partial^2 \phi}{\partial y^2} = \frac{1}{c^2} \frac{\partial^2 \phi}{\partial t^2}$$

such that ϕ vanishes on the lines $x = 0, x = a, y = 0, y = b$. Since the boundary conditions imposed on ϕ are such that the lines $x = 0, a$, and $y = 0, b$ are nodal lines, our solution must be of the stationary type. Referring to §**8**, equation (29), we see that possible solutions are

$$\phi = \frac{\cos}{\sin} px \frac{\cos}{\sin} qy \frac{\cos}{\sin} rct, \quad \text{where } p^2 + q^2 = r^2.$$

Since ϕ is identically zero at $x = 0$, and $y = 0$, we shall have to take the sine rather than the cosine in the first two factors. Further, since at $x = a, \phi = 0$ for all values of y, we must have

$$\sin pa = 0$$

and, similarly,

$$\sin qb = 0.$$

Hence $p = m\pi/a$, and $q = n\pi/b$, m and n being integers. A solution satisfying all the conditions is therefore

$$\phi = \sin (m\pi x/a) \sin (n\pi y/b) \genfrac{}{}{0pt}{}{\cos}{\sin} rct,$$

where

$$r^2 = \pi^2(m^2/a^2 + n^2/b^2).$$

The most general solution for any x, y in the rectangle $0 \le x \le a$, $0 \le y \le b$ and $t \ge 0$ is the sum of an arbitrary number of such terms,

$$\phi = \sum_{m,n} \sin (m\pi x/a) \sin (n\pi y/b)\{C_{mn} \cos rct + D_{mn} \sin rct\}. \quad (76)$$

To make this solution unique it is necessary to specify the constants C_{mn} and D_{mn} which requires knowledge of the initial conditions. Suppose ϕ and its derivative $\partial\phi/\partial t$ are known at $t = 0$, and that the infinite series (76) may be differentiated term by term. Then at $t = 0$, this gives

$$(\phi)_{t=0} = \sum C_{mn} \sin m\pi x/a \sin n\pi y/b,$$

$$(\partial\phi/\partial t)_{t=0} = \sum rcD_{mn} \sin m\pi x/a \sin n\pi y/b,$$

where the left-hand sides are known functions. Now by a suitable choice of the constants C_{mn} and D_{mn} we can always represent the known functions ϕ and $\partial\phi/\partial t$ at $t = 0$ in terms of these Fourier series representations. With this choice of C_{mn} and D_{mn} ϕ is then known for all $t \ge 0$ through (76). The precise method by which these constants may be found will be discussed later.

§16   Examples

1. Show that $\phi = f(x \cos \theta + y \sin \theta - ct)$ represents a wave in two dimensions, the direction of propagation making an angle θ with the axis of x.

2. Show that $\phi = a \cos (lx + my - ct)$ is a wave in two dimensions and find its wavelength.

3. What is the wavelength and speed of propagation of the system of plane waves $\phi = a \sin (Ax + By + Cz - Dt)$?

4. Show that three equivalent harmonic waves with 120° phase between each pair have zero sum.

5. Show that $\phi = r^{-1/2} \cos\frac{1}{2}\,\theta f(r \pm ct)$ is a progressive type wave in two dimensions, r and θ being plane polar coordinates, and f being an arbitrary twice differentiable function. By superposing two of these waves in which f is a harmonic function, obtain a stationary wave, and draw its nodal lines. Note that this is not a single-valued function unless we put restrictions upon the allowed range of θ.

6. By taking the special case of $f(x) = g(x) = \sin px$ in equation (24), show that it reduces to the result of equation (36) in which $m = n = 0$. Use the relation

$$J_{1/2}(x) = \left(\frac{2}{\pi x}\right)^{1/2} \sin x.$$

7. Find a solution of

$$\frac{\partial^2 \phi}{\partial x^2} + \frac{1}{c^2}\frac{\partial^2 \phi}{\partial t^2} = 0,$$

such that $\phi = 0$ when $t \to \infty$, and $\phi = 0$ when $x = 0$.

8. Find a solution of

$$\frac{\partial^2 \phi}{\partial x^2} = \frac{1}{c^2}\frac{\partial^2 \phi}{\partial t^2},$$

such that $\phi = 0$ when $x \to +\infty$ or $t \to +\infty$.

9. Solve the equation

$$\frac{\partial^2 z}{\partial t^2} = c^2 \frac{\partial^2 z}{\partial x^2},$$

given that z is never infinite for real values of x and t, and $z = 0$ when $x = 0$, or when $t = 0$.

10. Solve

$$\frac{\partial^2 V}{\partial x^2} = \frac{\partial V}{\partial t},$$

given that $V = 0$ when $t \to \infty$ and when $x = 0$, and when $x = l$.

11. x, y, z are given in terms of the three quantities ξ, η, ζ by the equations

$$x = a \sinh \xi \sin \eta \cos \zeta,$$

$$y = a \sinh \xi \sin \eta \sin \zeta,$$

$$z = a \cosh \xi \cos \eta.$$

Show that the equation

$$\frac{\partial^2 \phi}{\partial x^2} + \frac{\partial^2 \phi}{\partial y^2} + \frac{\partial^2 \phi}{\partial z^2} = \frac{1}{c^2} \frac{\partial^2 \phi}{\partial t^2}$$

is of the correct form for solution by the method of separation of variables, when ξ, η, ζ are used as the independent variables. Write down the subsidiary equations into which the whole equation breaks down.

12. Show that the equation of telegraphy (38) in one dimension has solutions of the form

$$\phi = \frac{\cos}{\sin} mx \frac{\cos}{\sin} rt \exp(-pt/2),$$

where m and r are constants satisfying the equation $r^2 = m^2 c^2 - \frac{1}{4} p^2$.

13. Consider the generalised equation of telegraphy in the form

$$\frac{\partial^2 \phi}{\partial x^2} = \frac{1}{c^2} \left\{ \frac{\partial^2 \phi}{\partial t^2} + (a+b) \frac{\partial \phi}{\partial t} + ab\phi \right\}.$$

By means of the substitution $\psi = \phi \exp\{\frac{1}{2}(a+b)t\}$ show that ψ satisfies the equation

$$\frac{\partial^2 \psi}{\partial t^2} - c^2 \frac{\partial^2 \psi}{\partial x^2} = \left(\frac{a-b}{2}\right)^2 \psi.$$

Hence deduce that propagation is relatively undistorted if $a = b$ and that the progressive wave solution in either direction is of the form

$$u = \exp(-at) f(x \pm ct),$$

with f an arbitrary twice differentiable function of its argument.

14. Find the solution to the inhomogeneous wave equation

$$\frac{\partial^2 \phi}{\partial t^2} - c^2 \frac{\partial^2 \phi}{\partial x^2} = \sin(kx - \omega t),$$

subject to the homogeneous initial conditions

$$\phi(x, 0) = 0 \quad \text{and} \quad \frac{\partial \phi}{\partial t}(x, 0) = 0.$$

Show that when $\omega/k \neq c$ the solution comprises three sinusoidal waves that propagate with constant but different amplitudes and with speeds $\pm c$ and ω/k. In the **resonance case**, when $\omega/k = c$, show that the solution comprises two constant amplitude harmonic waves that propagate with speeds $\pm c$ together with one harmonic wave whose amplitude grows linearly with time.

15. Complete the reasoning in §13 that gave rise to equation (68). Prove that $\partial \phi/\partial x$ and $\partial \phi/\partial t$ both satisfy corresponding expressions in relation to the corners A, B, C and D of a characteristic parallelogram.

16. Use the reasoning of §13 to show how d'Alembert's formula together with equation (68) can be used to solve the mixed initial and boundary value problem in the bounded region $0 \leqslant x \leqslant a$ and $t \geqslant 0$.

17. Consider the mixed initial and boundary value problem

$$\frac{\partial^2 \phi}{\partial t^2} - c^2 \frac{\partial^2 \phi}{\partial x^2} = 0,$$

subject to the initial conditions

$$\phi(x, 0) = h_1(x) \quad \text{and} \quad \frac{\partial \phi}{\partial t}(x, 0) = k_1(x) \quad \text{for} \quad x \geqslant 0$$

and the homogeneous free boundary condition

$$\frac{\partial \phi}{\partial x}(0, t) = 0.$$

By differentiating the general solution (52) partially with respect to x and using the free boundary condition, show that

$$f'(-x) = -g'(x).$$

Use this result, together with the equations that follow when equations (72) and (73) are differentiated with respect to x, to prove that if the solution is to be extended for negative x we must have

$$h_1'(-x) = -h_1'(x) \quad \text{and} \quad k_1(-x) = k_1(x).$$

Hence conclude that in the associated pure initial value problem the functions $h_1(x)$ and $k_1(x)$ must be extended for $x < 0$ as **even functions**, and that after *reflection* at $x = 0$ of the leftward moving wave its sign is unchanged.

18. Outline a method of approach using the notion of reflection that would enable the previous problem to be solved if the free boundary condition were to be replaced by a so-called **mixed boundary condition** of the form

$$a\phi(0, t) + b\frac{\partial\phi}{\partial x}(0, t) = 0.$$

Waves on strings

§17 The governing differential equation

In this chapter we shall discuss the transverse vibrations of a heavy string of mass ρ per unit length. By **transverse** vibrations we mean vibrations in which the displacement of each particle of the string is in a direction perpendicular to the length. When the displacement is in the same direction as the string, we call the waves **longitudinal**; these waves will be discussed in Chapter 4. We shall neglect the effect of gravity; in practice this may be approximately achieved by supposing that the whole motion takes place on a smooth horizontal plane.

In order that a wave may travel along the string, it is necessary that the string should be at least slightly extensible; in our calculations, however, we shall assume that the tension does not change appreciably from its normal value F. The condition for this (see **§18**) is that the wave disturbance is not too large.

Let us consider the motion of a small element of the string PQ (Fig. 5) of length δs. Suppose that in the equilibrium state the string lies along the axis of x, and that PQ is originally at $P_0 Q_0$. Let the displacement of PQ from the x axis be denoted by y. Then we shall obtain an equation for the motion of the element PQ in terms of the tension and density of the string. The forces acting on this element, when the string is vibrating, are merely the two tensions F acting along the tangents at P and Q as shown in the figure; let ψ and $\psi + \delta\psi$ be the angles made by these two tangents with the x axis. We can easily write down the equation of motion of the element PQ in the y direction; for the resultant force acting parallel to the y axis is $F \sin (\psi + \delta\psi) - F \sin \psi$. Neglecting squares of small quantities, this is $F \cos \psi \, \delta\psi$. The equation of motion is therefore

$$F \cos \psi \, \delta\psi = \rho \, \delta s \frac{\partial^2 y}{\partial t^2}. \tag{1}$$

Now we have

$$\tan \psi = \frac{\partial y}{\partial x}, \quad \text{so that} \sec^2 \psi \, \delta\psi = \frac{\partial^2 y}{\partial x^2} \, \delta x,$$

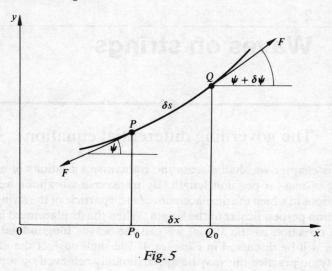

Fig. 5

so from (1)

$$\rho \frac{\partial^2 y}{\partial t^2} = F \cos^3 \psi \frac{\partial^2 y}{\partial x^2} \frac{\partial x}{\partial s}$$

$$= F \cos^4 \psi \frac{\partial^2 y}{\partial x^2}. \tag{2}$$

However, elementary arguments show

$$\cos^2 \psi = \left\{ 1 + \left(\frac{\partial y}{\partial x}\right)^2 \right\}^{-1},$$

when equation (2) becomes

$$\frac{\partial^2 y}{\partial t^2} = c^2 \left\{ 1 + \left(\frac{\partial y}{\partial x}\right)^2 \right\}^{-2} \frac{\partial^2 y}{\partial x^2}, \tag{3}$$

where $c^2 = F/\rho$.

This equation is nonlinear because of the presence of the bracketed term, but if the displacements involved are small enough for us to neglect $\left(\frac{\partial y}{\partial x}\right)^2$ compared with unity it may be *linearised* to give

$$\frac{\partial^2 y}{\partial x^2} = \frac{1}{c^2} \frac{\partial^2 y}{\partial t^2}. \tag{4}$$

We thus arrive at the linear wave equation already encountered in Chapter 1.

The general solution

$$y = f(x - ct) + g(x + ct)$$

to (4) represents waves of arbitrary shape moving in opposite directions along this infinite string, and with the constant speed $c = (F/\rho)^{1/2}$. The shape of these progressive waves will remain unchanged as they propagate.

A more complete discussion, in which we did not neglect terms of second order, would show us that the speed was not quite independent of the shape, and indeed, that the wave profile would change slowly with time. Something of this will be discussed later when in Chapter 9 we come to consider nonlinear equations, but for the time being we shall be content to work with (4). It is, indeed, an excellent approximation except when there is a sudden "kink" in y in which case we cannot neglect $\left(\dfrac{\partial y}{\partial x}\right)^2$.

§18   Kinetic and potential energies

Since the transverse speed of any point of the string is $\partial y/\partial t$ we can easily determine the kinetic energy of vibration. It is

$$T = \int \tfrac{1}{2}\rho\left(\frac{\partial y}{\partial t}\right)^2 dx. \tag{5}$$

The potential energy V is found by considering the increase of length of the element PQ. This element has increased its length from δx to δs. We have therefore done an amount of work $F(\delta s - \delta x)$. Summing for all the elements of the string, we obtain the formula

$$V = \int F(\delta s - \delta x) = \int F\left\{ \sqrt{\left(1 + \left(\frac{\partial y}{\partial x}\right)^2\right)} - 1 \right\} dx$$

$$= \tfrac{1}{2}F \int \left(\frac{\partial y}{\partial x}\right)^2 dx, \text{approximately.} \tag{6}$$

The integrations in (5) and (6) are both taken over the length of the string and they will be finite provided the disturbance is *localised*, for then the remainder of the string will be at rest.

With a localised progressive wave $y = f(x - ct)$ moving to the right with speed c, these equations give

$$T = \int \tfrac{1}{2}\rho c^2 (f')^2 \, dx = \tfrac{1}{2}F \int (f')^2 \, dx, \tag{7}$$

and

$$V = \tfrac{1}{2}F \int (f')^2 \, dx. \tag{8}$$

Thus the kinetic and potential energies are equal. The same result applies to the localised progressive wave $y = g(x + ct)$, moving to the left with speed c, but it does not, in general, apply to the stationary type waves $y = f(x - ct) + g(x + ct)$.

We can now decide whether our initial assumption is correct, that the tension remains effectively constant. If the string is elastic, the change in tension will be proportional to the change in length. We have seen in (6) that the change in length of an element δx is $\dfrac{1}{2}\left(\dfrac{\partial y}{\partial x}\right)^2 \delta x$.

Thus, provided that $\dfrac{\partial y}{\partial x}$ is of the first order of small quantities, the change of tension is of the second order, and may safely be neglected. This assumption is equivalent to asserting that the wave profile does not have any large "kinks", but has a relatively gradual variation with x.

The significance of the curvature of the string on the equation governing its motion can most easily be seen by observing that the radius of curvature r at P is, from elementary calculus,

$$r = \left\{ 1 + \left(\frac{\partial y}{\partial x}\right)^2 \right\}^{3/2} \bigg/ \left(\frac{\partial^2 y}{\partial x^2}\right),$$

which allows the right-hand side of (4) to be written $c^2 \cos \psi / r$.

§19   Inclusion of initial conditions

We may now employ d'Alembert's formula to obtain a solution to (4) once initial conditions are given for the string motion. These amount to specifying its initial shape and the initial transverse speed with which each point of this infinitely long string is moving.

So, setting

$$y(x, 0) = h(x) \text{ (initial shape)}$$

and

$$\frac{\partial y}{\partial t}(x, 0) = k(x), \text{(initial transverse speed)}$$

we at once deduce from equation (57), Chapter 1, that the subsequent displacement is given by

$$y(x, t) = \frac{h(x - ct) + h(x + ct)}{2} + \frac{1}{2c} \int_{x-ct}^{x+ct} k(s) \, ds. \tag{9}$$

§20   Reflection at a change of density

The discussion above applies specifically to strings of infinite length. Before we discuss strings of finite length, we shall solve two problems of reflection of waves from a *discontinuity* in the string. The first is when two strings of different densities are joined together, and the second is when a mass is concentrated at a point of the string. In each case we shall find that an incident wave gives rise to a *reflected* and a *transmitted* wave.

Consider first, then, the case of two semi-infinite strings 1 and 2 joined at the origin as in Fig. 6. Let the densities per unit length of the two strings be ρ_1 and ρ_2. Denote the displacements in the two strings by

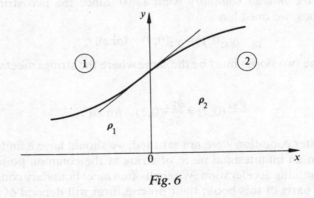

Fig. 6

y_1 and y_2. Let us suppose that a train of harmonic waves is incident from the negative direction of x. When these waves meet the change of string material, they will suffer partial reflection and partial transmission. If we choose the exponential functions of §10 to represent each of these waves, we may write

$$y_1 = y_{\text{incident}} + y_{\text{reflected}},$$

$$y_2 = y_{\text{transmitted}},$$

(10)

where

$$y_{\text{incident}} = A_1 \exp 2\pi i (nt - k_1 x),$$

$$y_{\text{reflected}} = B_1 \exp 2\pi i (nt + k_1 x),$$

(11)

$$y_{\text{transmitted}} = A_2 \exp 2\pi i (nt - k_2 x).$$

A_1 is real, but A_2 and B_1 may be complex. According to §10 equation (42), the arguments of A_2 and B_1 will give their phases relative to the incident wave. All three waves in (11) must have the same frequency n, but since the wave speeds in the two wires are different, they will have different wavelengths $1/k_1$ and $1/k_2$. The reflected wave must, of course, have the same wavelength as the incident wave. Since the velocities of the two types of waves are n/k_1 and n/k_2 (Chapter 1, equations (7) and (10)), and we have shown in (3) that $c^2 = F/\rho$, therefore

$$k_1^2 / k_2^2 = \rho_1 / \rho_2.$$

(12)

In order to determine A_2 and B_1 we must now use the appropriate boundary conditions. In this case these are the conditions which must hold at the *interior boundary point* $x = 0$. Since the two strings are continuous, we must have

$$y_1(0, t) = y_2(0, t) \quad \text{for all } t,$$

and as the two slopes must be the same where the strings meet we also require

$$\frac{\partial y_1}{\partial x}(0, t) = \frac{\partial y_2}{\partial x}(0, t) \quad \text{for all } t.$$

If this latter condition were not satisfied, we should have a finite force acting on an infinitesimal piece of string at the common point, thus giving it infinite acceleration. We shall often meet boundary conditions in other parts of this book; their precise form will depend of course upon the particular problem under discussion. In the present case, the two boundary conditions give

$$A_1 + B_1 = A_2,$$

$$2\pi i(-k_1 A_1 + k_1 B_1) = 2\pi i(-k_2 A_2).$$

These equations have a solution

$$\frac{B_1}{A_1} = \frac{k_1 - k_2}{k_1 + k_2}, \qquad \frac{A_2}{A_1} = \frac{2k_1}{k_1 + k_2}. \tag{13}$$

Since k_1, k_2 and A_1 are real, this shows that B_1 and A_2 are both real. A_2 is positive for all k_1 and k_2, but B_1 is positive if $k_1 > k_2$, and negative if $k_1 < k_2$. Thus the transmitted wave is always in phase with the incident wave, but the reflected wave is in phase only when the incident wave is in the denser medium; otherwise it is exactly out of phase.

The **coefficient of reflection** R is defined to be the ratio, $R = |B_1/A_1|^2 = \left(\dfrac{k_1 - k_2}{k_1 + k_2}\right)^2$ which, by (12), we may write

$$\left(\frac{\sqrt{\rho_1} - \sqrt{\rho_2}}{\sqrt{\rho_1} + \sqrt{\rho_2}}\right)^2. \tag{14}$$

Since, from (7) and (8), the energy of a progressive wave is proportional to the square of its amplitude, it follows that R represents the ratio of reflected energy to incident energy. Similarly, since no energy is wasted, the coefficient of transmission T, which gives the ratio of transmitted energy to incident energy, is equal to $1 - R$,

$$T = \frac{4\sqrt{\rho_1}\sqrt{\rho_2}}{(\sqrt{\rho_1} + \sqrt{\rho_2})^2}. \tag{15}$$

§21   Reflection at a concentrated load

A similar discussion can be given for the case of a mass M concentrated at a point of the string. Let us take the equilibrium position of the mass to be the origin as in Fig. 7, and suppose that the string is identical on the two sides. Then if the incident wave comes from the negative side of the origin, we may write, just as in (10) and (11):

$$y_1 = y_{\text{incident}} + y_{\text{reflected}}$$

$$y_2 = y_{\text{transmitted}}$$

where

$$y_{\text{incident}} = A_1 \exp 2\pi i(nt - kx),$$

$$y_{\text{reflected}} = B_1 \exp 2\pi i(nt + kx), \tag{16}$$

$$y_{\text{transmitted}} = A_2 \exp 2\pi i(nt - kx).$$

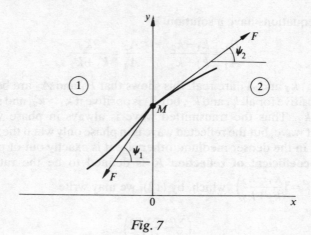

Fig. 7

The boundary conditions are that for all values of t

(i) $y_1(0, t) = y_2(0, t)$ (17)

(ii) $F\left[\dfrac{\partial y_2}{\partial x}(0, t) - \dfrac{\partial y_1}{\partial x}(0, t)\right] = M\dfrac{\partial^2 y_2}{\partial t^2}$. (18)

The first equation expresses the *continuity* of the string and the second is the *equation of motion* of the mass M. We can see this as follows: the net force on m is the difference of the components of F on either side, so that if ψ_1 and ψ_2 are the angles made by tangents to the string at M with the x axis, we have

$$M\frac{\partial^2 y_2}{\partial t^2}(0, t) = F(\sin \psi_2 - \sin \psi_1).$$

Since ψ_1 and ψ_2 are assumed to be small, we may put $\sin \psi_2 = \tan \psi_2 = \dfrac{\partial y_2}{\partial x}$, $\sin \psi_1 = \dfrac{\partial y_1}{\partial x}$, and (18) is then obtained.

Substituting from (16) into (17) and (18), and cancelling the term $\exp 2\pi i n t$, which is common to both sides, we find

$$A_1 + B_1 = A_2,$$

$$2\pi i k F(A_2 - A_1 + B_1) = 4\pi^2 n^2 M A_2.$$

Let us write

$$\pi n^2 M / k F = p.$$ (19)

Solving these equations then gives

$$\frac{B_1}{A_1} = \frac{-ip}{1+ip} = \frac{-p^2 - ip}{1+p^2}, \tag{20}$$

$$\frac{A_2}{A_1} = \frac{1}{1+ip} = \frac{1-ip}{1+p^2}. \tag{21}$$

In this problem, unlike the last, B_1 and A_2 are complex, so that there are phase changes. These phases (according to §10) are given by the arguments of (20) and (21). They are therefore $\tan^{-1}(p)$ and $\tan^{-1}(-1/p)$, respectively. The coefficient of reflection $R = |B_1/A_1|^2$, which equals $p^2/(1+p^2)$, and the coefficient of transmission T is $1 - R = 1/(1+p^2)$. If we write $p = \tan\theta$, where $0 \le \theta \le \pi/2$, then we find that the phase changes are θ and $\pi/2 + \theta$, and also $R = \sin^2\theta$, $T = \cos^2\theta$.

§22   Alternative solutions

The two problems in §§20, 21 could be solved quite easily by taking a real form for each of the waves instead of the complex forms (11) and (16). The reader is advised to solve these problems in this way, taking, for example, in §21, the forms

$$y_{\text{incident}} = a_1 \cos 2\pi(nt - kx),$$

$$y_{\text{reflected}} = b_1 \cos\{2\pi(nt + kx) + \varepsilon\}, \tag{22}$$

$$y_{\text{transmitted}} = a_2 \cos\{2\pi(nt - kx) + \eta\}.$$

In most cases of progressive waves, however, the complex form is the easier to handle; the reason for this is that exponentials are simpler than harmonic functions, and also the amplitude and phase are represented by one complex quantity rather than by two separate terms.

§23   Strings of finite length, normal modes

So far we have been dealing with strings of infinite length. When we deal with strings of finite length it is easier to use stationary type waves instead of progressive type waves. Let us now consider waves on a string of length l, fastened at the ends where $x = 0, l$. We have to find a solution of the wave equation

$$\frac{\partial^2 y}{\partial x^2} = \frac{1}{c^2} \frac{\partial^2 y}{\partial t^2},$$

subject to the boundary conditions $y = 0$, at $x = 0, l$, for all t. Now by Chapter 1, §8, we see that suitable solutions are of the type

$$\frac{\cos}{\sin} px \frac{\cos}{\sin} cpt.$$

It is clear that the cosine term in x will not satisfy the boundary condition at $x = 0$, and we may therefore write the solution

$$y = \sin px \, (a \cos cpt + b \sin cpt).$$

The constants a, b and p are arbitrary, but we have still to make $y = 0$ at $x = l$. This implies that $\sin pl = 0$, or that $pl = \pi, 2\pi, 3\pi \ldots$. It follows that the solution is

$$y = \sin \frac{r\pi x}{l} \left(a \cos \frac{r\pi ct}{l} + b \sin \frac{r\pi ct}{l} \right), r = 1, 2, 3, \ldots. \qquad (23)$$

Each of the solutions (23), in which r may have any positive integral value, is known as a **normal mode** of vibration. It is also called an **eigenfunction** of the wave equation corresponding to the given boundary conditions. The most general solution is the sum of any number of terms similar to (23) and may therefore be written

$$y = \sum_r \sin \frac{r\pi x}{l} \left\{ a_r \cos \frac{r\pi ct}{l} + b_r \sin \frac{r\pi ct}{l} \right\}. \qquad (24)$$

The values of a_r and b_r are determined by means of the initial conditions

$$y(x, 0) = h(x) \text{ (initial shape)}$$

and

$$\frac{\partial y}{\partial t}(x, 0) = k(x) \text{ (initial transverse speed)}.$$

Thus, when $t = 0$

$$y(x, 0) = h(x) = \sum_r a_r \sin r\pi x/l, \qquad (25)$$

and

$$\frac{\partial y}{\partial t}(x, 0) = k(x) = \sum_r b_r (r\pi c/l) \sin r\pi x/l. \tag{26}$$

Once h and k have been specified, then each a_r and b_r is found from (25) and (26), and hence the full solution is obtained. We shall write down the results for reference. The Fourier analysis represented by (25) and (26) gives

$$a_r = \frac{2}{l} \int_0^l h(x) \sin \frac{r\pi x}{l} \, dx$$

$$b_r = \frac{2}{r\pi c} \int_0^l k(x) \sin \frac{r\pi x}{l} \, dx. \tag{27}$$

In particular, if the string is released from rest when $t = 0$, every $b_r = 0$.

§24   String plucked at mid-point

As an illustration of the theory of the last section, let us consider the case of a plucked string of length l released from rest when the mid-point is drawn aside through a distance h (as in Fig. 8). In

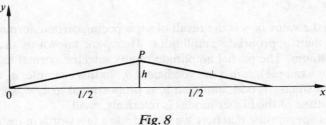

Fig. 8

accordance with (25) and (26) we can assume that

$$y = \sum_{r=1}^{\infty} a_r \sin \frac{r\pi x}{l} \cos \frac{r\pi c t}{l}$$

When $t = 0$, this reduces to $\sum_r a_r \sin \frac{r\pi x}{l}$, and the coefficients a_r have to be chosen so that this is identical with

$$y = \frac{2h}{l}x, \qquad 0 \leqslant x \leqslant \tfrac{1}{2}l$$

$$y = \frac{2h}{l}(l-x), \qquad \tfrac{1}{2}l \leqslant x \leqslant l.$$

If we multiply both sides of the equation

$$y = \sum_r a_r \sin\frac{r\pi x}{l}$$

by $\sin r\pi x/l$ and integrate from $x = 0$ to $x = l$, as in the method of Fourier analysis, all the terms except one will disappear on the right-hand side, and we shall obtain

$$\frac{l}{2}a_r = \int_0^{l/2} \frac{2h}{l}x \sin\frac{r\pi x}{l}\,\mathrm{d}x + \int_{l/2}^{l} \frac{2h}{l}(l-x) \sin\frac{r\pi x}{l}\,\mathrm{d}x.$$

Whence

$$a_r = \frac{8h}{\pi^2 r^2}\sin\frac{r\pi}{2} \text{ when } r \text{ is odd,}$$

$$= 0 \text{ when } r \text{ is even.}$$

So the full solution, giving the value of y at all subsequent times, is

$$y = \frac{8h}{\pi^2} \sum_{n=0}^{\infty} \frac{1}{(2n+1)^2}\sin\frac{(2n+1)\pi}{2}\sin\frac{(2n+1)\pi x}{l}\cos\frac{(2n+1)\pi ct}{l} \quad (28)$$

Thus the value of y is the result of superposing certain normal modes with their appropriate amplitudes. These are known as the **partial amplitudes**. The partial amplitude of any selected normal mode (the rth for example), is just the coefficient a_r. In this example, a_r vanishes except when r is odd, and then a_r is proportional to $1/r^2$, so that the amplitude of the higher modes is relatively small.

It is appropriate that here we should add a few words in justification of this example, since it apparently violates various assumptions already made. Namely, that the initial data should be assumed to be differentiable and that the string should have no sudden "kinks" in it. Both of these conditions are violated at point P in Fig. 8.

The justification for our analysis comes, in fact, from the result proved at the end of §11 and from the assumption that the displacement is small. The first result enables us to replace the initial profile in Fig. 8 by a smooth approximation to it by rounding off the kink at P. The assumption of a small displacement ensures that the radius of curvature of the approximation to the string profile at P is not unreasonably small, so that equation (4) rather than (3) is the appropriate one with which to work.

§25 Energies of normal modes

The rth normal mode (23) has a frequency $rc/2l$. Also, there are zero values of y (i.e. nodes) at the points $x = 0, l/r, 2l/r, \ldots, (r-1)l/r, l$. If the string is plucked with the finger lightly resting on the point l/r it will be found that this mode of vibration is excited. With even-order vibrations (r even) the mid-point is a node, and with odd-order vibrations it is an antinode.

We can find the energy associated with this mode of vibration most conveniently by rewriting (23) in the form

$$y = A \sin \frac{r\pi x}{l} \cos \left\{ \frac{r\pi ct}{l} + \varepsilon \right\}. \tag{29}$$

Here A is the amplitude and ε is the phase. According to (5) the kinetic energy is

$$T = \tfrac{1}{2}\rho \int_0^l \left(\frac{\partial y}{\partial t} \right)^2 \mathrm{d}x = \frac{\pi^2 c^2 r^2 \rho}{4l} A^2 \sin^2 \left\{ \frac{r\pi ct}{l} + \varepsilon \right\}. \tag{30}$$

Similarly, by (6) the potential energy is

$$V = \tfrac{1}{2}F \int_0^l \left(\frac{\partial y}{\partial x} \right)^2 \mathrm{d}x = \frac{\pi^2 r^2 F}{4l} A^2 \cos^2 \left\{ \frac{r\pi ct}{l} + \varepsilon \right\}. \tag{31}$$

Now $F/\rho = c^2$, and so the two coefficients in (30) and (31) are equal. The total energy of this vibration is therefore

$$\frac{\pi^2 c^2 r^2 \rho}{4l} A^2. \tag{32}$$

The total energy is thus proportional to the square of the amplitude and also to the square of the frequency. This is a result that we shall often find as we investigate various types of wave motion.

As a rule, however, there are several normal modes present at the same time, and we can then write the displacement (24) in the more convenient form

$$y = \sum_{r=1}^{\infty} A_r \sin \frac{r\pi x}{l} \cos \left\{ \frac{r\pi ct}{l} + \varepsilon_r \right\}. \tag{33}$$

A_r is the amplitude, and ε_r is the phase, of the rth normal mode. When we evaluate the kinetic energy as in (30) we find that the "cross-terms" vanish, since

$$\int_0^l \sin\frac{r\pi x}{l} \sin\frac{s\pi x}{l}\,dx = 0, \quad \text{if } r \neq s.$$

Consequently the total kinetic energy is just

$$\frac{\pi^2 c^2 \rho}{4l} \sum r^2 A_r^2 \sin^2\left\{\frac{r\pi ct}{l}+\varepsilon_r\right\},$$

and in a precisely similar way the total potential energy is

$$\frac{\pi^2 F}{4l} \sum r^2 A_r^2 \cos^2\left\{\frac{r\pi ct}{l}+\varepsilon_r\right\}.$$

By addition we find that the total energy of vibration is

$$\frac{\pi^2 c^2 \rho}{4l} \sum r^2 A_r^2. \tag{34}$$

This formula is important. It shows that the total energy is merely the sum of the energies obtained separately for each normal mode. It is due to this simple fact, which arises because there are no cross-terms involving $A_r A_s$, that the separate modes of vibration are called *normal modes*. It should be observed that this result holds for both the kinetic and potential energies separately as well as for their sum.

We have already seen that when a string vibrates, more than one mode is usually excited. The lowest frequency, $c/2l$, is called the **ground note**, or **fundamental**, and the others, with frequencies $rc/2l$, are **harmonics** or **overtones**. The frequency of the fundamental varies directly as the square root of the density. This is known as **Mersenne's law**. The **tone**, or quality, of a vibration is governed by the proportion of energy in each of the harmonics, and it is this that is characteristic of each musical instrument. The tone must be carefully distinguished from the pitch, which is merely the frequency of the fundamental.

We can use the results of (34) to determine the total energy in each normal mode of the vibrating string which we discussed in §24. According to (28) and (33) $A_{2n} = 0$, and

$$A_{2n+1} = \frac{8h}{\pi^2}\frac{1}{(2n+1)^2}\sin\frac{(2n+1)\pi}{2}.$$

Consequently, the total energy of the even modes is zero, and the energy of the $(2n+1)$th mode is $16c^2h^2\rho/(2n+1)^2\pi^2l$. This shows us that the main part of the energy is associated with the normal modes of low order. We can check these formulae for the energies in this example quite easily. For the total energy of the whole vibration is the sum of the energies of each normal mode separately:

$$\text{total energy} = \frac{16c^2h^2\rho}{\pi^2l} \sum_{n=0}^{\infty} \frac{1}{(2n+1)^2}.$$

Now $1/1^2+1/3^2+1/5^2+\ldots = \pi^2/8$ so the total energy is $2c^2h^2\rho/l$, or $2Fh^2/l$. But the string was drawn aside and released from rest in the position of Fig. 8, and at that moment the whole energy was in the form of potential energy. This potential energy is just F times the increase in length, $2F\{(l^2/4+h^2)^{1/2}-l/2\}$. A simple calculation shows that if we neglect powers of h above the second, as we have already done in our formulation of the equation of wave motion, this becomes $2Fh^2/l$, thus verifying our earlier result.

This particular example corresponds quite closely to the case of a violin string bowed at its mid-point. A listener would thus hear not only the fundamental, but also a variety of other frequencies, simply related to the fundamental numerically. This would not therefore be a pure note, though the small amount of the higher harmonics makes it much purer than that of many musical instruments, particularly a piano.

If the string had been bowed at some other point than its centre, the partial amplitudes would have been different, and thus the tone would be changed. By choosing the point properly any desired harmonic may be emphasised or diminished, a fact well known to musicians.

§26   Normal coordinates

We have seen in §25 that it is most convenient to analyse the motion of a string of finite length in terms of its normal modes. According to (33) the rth **mode** of the rth **eigenfunction** is

$$y_r = A_r \sin\frac{r\pi x}{l} \cos\left\{\frac{r\pi ct}{l}+\varepsilon_r\right\}.$$

We often write this

$$y_r = \phi_r \sin\frac{r\pi x}{l}. \tag{35}$$

The expressions ϕ_r are known as the **normal coordinates** for the string. There are an infinite number of these coordinates, since there are an infinite number of degrees of freedom in a vibrating string. The advantage of using these coordinates can be seen from (30) and (31); if the displacement of the string is

$$y = \sum_{r=1}^{\infty} \phi_r \sin \frac{r\pi x}{l}. \tag{36}$$

then

$$T = \tfrac{1}{4}\rho l \sum_r \dot{\phi}_r^2,$$

$$V = \frac{\pi^2 c^2 \rho}{4l} \sum_r r^2 \phi_r^2, \tag{37}$$

where a dot denotes differentiation with respect to time.

The reason why we call ϕ_r a normal coordinate is now clear; for in mechanics the normal coordinates $q_1, q_2 \ldots q_n$ are suitable combinations of the original variables so that the kinetic and potential energies can be written in the form

$$T = a_1 \dot{q}_1^2 + a_2 \dot{q}^2 + a_3 \dot{q}_3^2 + \ldots,$$

$$V = b_1 q_1^2 + b_2 q_2^2 + b_3 q_3^2 + \ldots. \tag{38}$$

The similarity between (37) and (38) is obvious. Further, it can be shown, though we shall not reproduce the analysis here, that Lagrange's equations of motion apply with the set of coordinates ϕ_r in just the same way as with the coordinates q_r in ordinary mechanics.

§27 String with load at its mid-point

We shall next discuss the normal modes of a string of length l when a mass M is tied to its mid-point as in Fig. 9. Now we have already seen in §25 that in the normal vibrations of an unloaded string the normal modes of even order have a node at the mid-point. In such a vibration there is no motion at this point, and it is clearly irrelevant whether there is or is not a mass concentrated there. Accordingly, the normal modes or even order are unaffected by the presence of the mass, and our discussion will apply to the odd normal modes.

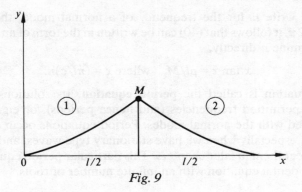

Fig. 9

Just as in the calculations of §§**20, 21**, in which there was a discontinuity in the string, we shall have two separate expressions y_1 and y_2 valid in the regions $0 \leqslant x \leqslant l/2$ and $l/2 \leqslant x \leqslant l$. It is obvious that the two expressions must be such that y is symmetrical about the mid-point of the string. y_1 must vanish at $x = 0$ and y_2 at $x = l$. Consequently, we may try the solutions

$$y_1 = a \sin px \cos (cpt + \varepsilon),$$

$$y_2 = a \sin p(l - x) \cos (cpt + \varepsilon). \tag{39}$$

We have already satisfied the boundary condition $y_1 = y_2$ at $x = l/2$. There is still the other boundary condition which arises from the motion of M. Just as in (18) we may write this

$$F\left[\frac{\partial y_2}{\partial x}(l/2, t) - \frac{\partial y_1}{\partial x}(l/2, t)\right] = M\frac{\partial^2 y_2}{\partial t^2}(l/2, t).$$

Substituting the values of y_1 and y_2 as given by (39) and using the relation $F = c^2\rho$, we find

$$\frac{pl}{2} \tan \frac{pl}{2} = \frac{\rho l}{M} = \text{const.} \tag{40}$$

The quantity $pl/2$ is therefore any one of the roots of the equation $x \tan x = \rho l/M$. If we draw the curves $y = \tan x$, $y = \rho l/Mx$, we can see that these roots lie in the regions 0 to $\pi/2$, π to $3\pi/2$, 2π to $5\pi/2$, etc. If we call the roots $x_1, x_2 \ldots$ then the frequencies $cp/2\pi$ become $cx_r/\pi l$. If M is zero so that the string is unloaded, $x_r = (r + 1/2)\pi$, so the presence of M has the effect of decreasing the frequencies of odd order.

If we write n for the frequency of a normal mode, then, since $n = cp/2\pi$, it follows that (40) can be written in the form of an equation to determine n directly,

$$x \tan x = \rho l/M, \quad \text{where } x = (\pi l/c)n. \tag{41}$$

This equation is called the **period equation**. Its solutions are the various permitted frequencies (and hence periods), or eigenvalues, associated with the normal modes. Period equations occur very frequently, especially when we have stationary type waves, and we shall often meet them in later chapters. This particular period equation is a transcendental equation with an infinite number of roots.

§28   Damped vibrations

In the previous paragraphs we have assumed that there was no frictional resistance, so that the vibrations were undamped. In practice, however, the air does provide a resistance to motion; this is roughly proportional to the velocity. Let us therefore discuss the motion of a string of length l fixed at its ends but subject to a resistance proportional to the velocity. The fundamental equation of wave motion (4) has to be supplemented by a term in $\dfrac{\partial y}{\partial t}$ and it becomes

$$\frac{\partial^2 y}{\partial x^2} = \frac{1}{c^2}\left\{\frac{\partial^2 y}{\partial t^2} + p\frac{\partial y}{\partial t}\right\}. \tag{42}$$

A solution by the method of separation of variables (cf. §9) is easily obtained, and we find

$$y = A \exp\left(-\tfrac{1}{2}pt\right)\sin \alpha x \cos\left(\sqrt{(c^2\alpha^2 - p^2/4)}t + \varepsilon\right).$$

Since y is to vanish at the two ends, we must have, as before, $\sin \alpha l = 0$, and hence $\alpha = r\pi/l, r = 1, 2, 3, \ldots$. The normal modes of vibration are therefore

$$y = A_r \exp\left(-\tfrac{1}{2}pt\right)\sin\frac{r\pi x}{l}\cos\left(qt + \varepsilon_r\right), \tag{43}$$

where

$$q^2 = \frac{r^2\pi^2 c^2}{l^2} - \frac{p^2}{4}.$$

The exponential term $\exp\left(-\tfrac{1}{2}pt\right)$ represents a decaying amplitude with modulus (see §9) equal to $2/p$. The frequency $q/2\pi$ is slightly less than when there is no frictional resistance. However, p is usually small, so that this decrease in frequency is often so small that it may be neglected.

§29   Method of reduction to a steady wave

There is another interesting method of obtaining the speed of propagation of waves along a string, which we shall now describe and which is known as the method of **reduction to a steady wave**. Suppose that a wave is moving from left to right in Fig. 10 with speed c. Then, if we

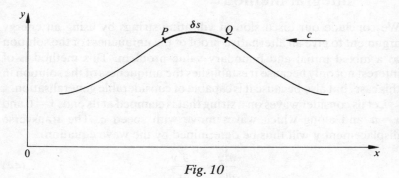

Fig. 10

superimpose on the whole motion a uniform speed $-c$ the wave profile itself will be reduced to rest, and the string will everywhere be moving with speed c, keeping all the time to a fixed curve (the wave profile). We are thus led to a different problem from our original one; for now the string is moving and the wave profile is at rest, whereas originally the wave profile was moving and the string as a whole was at rest. Consider the motion of the small element PQ of length δs situated at the top of the hump of a wave. If r is the radius of curvature at this top point, and we suppose, as in §17, that the string is almost inextensible, then the acceleration of the element PQ is c^2/r downwards. Consequently, the forces acting on it must reduce to $(c^2/r)\rho\,\delta s$. But these forces are merely the two tensions F at P and Q and, just as in §17, they give a resultant $F\,\delta s/r$ downwards. Equating the two expressions, we have

$$\frac{c^2\rho}{r}\,\delta s = F\frac{\delta s}{r} \quad \text{so that } c^2 = F/\rho.$$

This is, naturally, the same result as found before. The disadvantage of this method is that it does not describe in detail the propagation of the wave, nor does it deal with stationary waves, so that we cannot use it to get the equation of wave motion, etc. It is, however, very useful if we are only concerned with the wave speed, and we shall see later that this simple artifice of reducing the wave to rest can be used in other problems as well.

§30   Uniqueness of motion by the energy integral method

We conclude our discussion of vibrating strings by using an energy argument to give an alternative proof of the uniqueness of the solution to a mixed initial and boundary value problem. This method is of interest not only because it establishes the uniqueness of the solution in this case, but also because it is capable of considerable generalisation.

Let us consider waves on a string that is clamped at its ends $x = 0$ and $x = a$ and along which waves move with speed c. The transverse displacement y will thus be determined by the wave equation.

$$\frac{\partial^2 y}{\partial t^2} = c^2 \frac{\partial^2 y}{\partial x^2}, \tag{44}$$

for which the boundary conditions are

$$y(0, t) = y(a, t) = 0 \quad \text{for all } t. \tag{45}$$

For initial conditions we will assume that

$$y(x, 0) = h(x) \quad \text{and} \quad \frac{\partial y}{\partial t}(x, 0) = k(x), \tag{46}$$

where h, k are arbitrary functions with the usual differentiability properties and subject only to the conditions $h(0) = h(a) = k(0) = k(a) = 0$ in order to be compatible with the boundary conditions.

Suppose, if possible, that two different solutions y_1 and y_2 exist satisfying these mixed initial and boundary conditions and set $w = y_1 - y_2$. Then, because of the linearity of the wave equation, we have

$$\frac{\partial^2 w}{\partial t^2} = c^2 \frac{\partial^2 w}{\partial x^2}, \tag{47}$$

while from its manner of definition w must satisfy the homogeneous boundary conditions

$$w(0, t) = w(a, t) = 0 \quad \text{for all } t, \tag{48}$$

and the homogeneous initial conditions

$$w(x, 0) = 0 \quad \text{and} \quad \frac{\partial w}{\partial t}(x, 0) = 0 \quad \text{for} \quad 0 \leqslant x \leqslant a. \tag{49}$$

From (5) and (6) we know that the total energy $E(t)$ at time t is

$$E(t) = \int_0^a \left\{ \frac{1}{2}\rho\left(\frac{\partial w}{\partial t}\right)^2 + \frac{1}{2}F\left(\frac{\partial w}{\partial x}\right)^2 \right\} \, \mathrm{d}x, \tag{50}$$

so that as $c^2 = F/\rho$,

$$\frac{\mathrm{d}E}{\mathrm{d}t} = \rho \int_0^a \left\{ \frac{\partial w}{\partial t}\frac{\partial^2 w}{\partial t^2} + c^2 \frac{\partial w}{\partial x}\frac{\partial^2 w}{\partial x \, \partial t} \right\} \, \mathrm{d}x. \tag{51}$$

Using (47) to replace $\partial^2 w/\partial t^2$ in the integral in (51) then enables us to write

$$\frac{\mathrm{d}E}{\mathrm{d}t} = \rho c^2 \int_0^a \frac{\partial}{\partial x}\left(\frac{\partial w}{\partial t}\frac{\partial w}{\partial x}\right) \, \mathrm{d}x,$$

or

$$\frac{\mathrm{d}E}{\mathrm{d}t} = \rho c^2 \left\{ \left(\frac{\partial w}{\partial t}\frac{\partial w}{\partial x}\right)_{x=0} - \left(\frac{\partial w}{\partial t}\frac{\partial w}{\partial x}\right)_{x=a} \right\}.$$

This result taken together with the second initial condition in (49) used for $x = 0, a$ then shows

$$\frac{\mathrm{d}E}{\mathrm{d}t} = 0 \quad \text{so that } E(t) = \text{const.} \tag{52}$$

Setting $t = 0$ in (50) and employing both conditions in (49), the first after differentiation with respect to x, gives the result

$$E(t) \equiv 0 \quad \text{for all } t. \tag{53}$$

However, as the integrand of (50) is essentially non-negative this implies at once that

$$\frac{\partial w}{\partial t}(x, t) = 0 \quad \text{and} \quad \frac{\partial w}{\partial x}(x, t) = 0,$$

or

$$w(x, t) = \text{const,}$$

for $0 \le x \le a$ and all $t \ge 0$. As $w = 0$ on the initial line we must have $w = y_1 - y_2 \equiv 0$ for $0 \le x \le a$ and all $t \ge 0$. This proves that the solution is unique.

It follows as a direct corollary of this result that the quantity

$$E(t) = \int_0^a \left\{ \frac{1}{2}\rho\left(\frac{\partial y}{\partial t}\right)^2 + \frac{1}{2}F\left(\frac{\partial y}{\partial x}\right)^2 \right\} dx \qquad (54)$$

is an **invariant** of the mixed initial and boundary value problem (47), (48) and (49). Expressed differently, we may say that the quantity $E(t)$ is **conserved** in such a problem. This is a strict mathematical consequence of the form of the problem and involves no approximation. Arguments employing conservation of energy have, of course, already been used elsewhere in this chapter, though there the justification for them was essentially based on physical grounds. It only becomes necessary to make any approximation when $E(t)$ is identified with the energy of wave motion on a stretched string. For obvious reasons, arguments involving an expression analogous to (54) are called **energy integral** methods.

§31   Examples

1. Find the speed of waves along a string whose density per unit length is $0 \cdot 4$ kg m^{-1} when stretched to a tension $0 \cdot 9$ N.

2. A string of unlimited length is pulled into a harmonic shape $y = a \cos kx$, and at time $t = 0$ it is released. Show that if F is the tension and ρ the density per unit length of the string, its shape at any subsequent time t is $y = a \cos kx \cos kct$, where $c^2 = F/\rho$. Find the mean kinetic and potential energies per unit length of string.

3. Find the reflection coefficient for two strings which are joined together and whose line densities are $2 \cdot 5$ kg m^{-1} and $0 \cdot 9$ kg m^{-1}.

4. An infinite string lies along the x axis. At $t = 0$ that part of it between $x = \pm a$ is given a transverse velocity $a^2 - x^2$. Describe, with the help of equation (9), the subsequent motion of the string, the speed of wave motion being c.

5. Investigate the same problem as in question (4) except that the string is finite and of length $2a$, fastened at the points $x = \pm a$.

6. What is the total energy of the various normal modes in question (5)? Verify, by summation over all the normal modes, that this is equal to the initial kinetic energy.

7. The two ends of a uniform stretched string are fastened to light rings that can slide freely on two fixed parallel wires $x = 0, x = l$, one ring being on each wire. Find the normal modes of vibration.

8. A uniform string of length $3l$ fastened at its ends, is plucked a distance a at a point of trisection. It is then released from rest. Find the energy in each of the normal modes and verify that the sum is indeed equal to the work done in plucking the string originally.

9. Discuss fully the period equation (41) in §27. Show in particular that successive values of x approximate to $r\pi$, and that a closer approximation is $x = r\pi + \rho l/Mr\pi$.

10. Show that the total energy of vibration (43) is

$$\tfrac{1}{4}\rho l A_r^2 \exp{(-pt)}\{q^2 + pq \cos{(qt + \varepsilon_r)} \sin{(qt + \varepsilon_r)} + \tfrac{1}{2}p^2 \cos^2{(qt + \varepsilon_r)}\},$$

and hence prove that the rate of dissipation of energy is

$$\tfrac{1}{8}p\rho l A_r^2 \exp{(-pt)}\{2q \sin{(qt + \varepsilon_r)} + p \cos{(qt + \varepsilon_r)}\}^2.$$

11. Two uniform wires of densities ρ_1 and ρ_2 per unit length and of equal length are fastened together at one end and the other two ends are tied to two fixed points a distance $2l$ apart. The tension is F. Find the normal periods of vibration.

12. The density per unit length of a stretched string is m/x^2. The endpoints are at $x = a, 2a$, and the tension is F. Verify that the normal vibrations are given by the expression

$$y = A \sin{[\theta \log_e{(x/a)}]}\left(\frac{x}{a}\right)^{1/2} \frac{\cos}{\sin} pt, \text{ where } \theta^2 = \frac{mp^2}{F} - \frac{1}{4}.$$

Show that the period equation is $\theta \log_e 2 = n\pi, n = 1, 2, \ldots$

13. A heavy uniform chain of length l hangs freely from one end, and performs small lateral vibrations. Show that the normal vibrations are given by the expression

$$y = A J_0(2p\sqrt{\{x/g\}}) \cos{(pt + \varepsilon)},$$

where J_0 represents Bessel's function (§7) of order zero, x being measured from the lower end.

Deduce that the period equation is $J_0(2p\sqrt{\{l/g\}}) = 0$.

14. Use the method of §30 to prove the uniqueness of the mixed initial and boundary value problem

$$\frac{\partial^2 y}{\partial t^2} = c^2 \frac{\partial^2 y}{\partial x^2},$$

subject to the initial conditions

$$y(x, 0) = h(x) \quad \text{and} \quad \frac{\partial y}{\partial t}(x, 0) = k(x),$$

and the mixed boundary conditions

$$\alpha y(0, t) + \beta \frac{\partial y}{\partial x}(0, t) = 0$$

and

$$\alpha y(a, t) + \beta \frac{\partial y}{\partial x}(a, t) = 0,$$

for all $t \geq 0$, where α, β are constants.

3
Waves in membranes

§32 The governing differential equation

The vibrations of a plane membrane stretched to a uniform tension T may be discussed in a manner very similar to that which we have used in Chapter 2 for strings. When we say that the tension is T we mean that if a line of unit length is drawn in the surface of the membrane, then the material on one side of this line exerts a force T on the material on the other side and this force is perpendicular to the line we have drawn. Let us consider the vibrations of such a membrane; we shall suppose that its thickness may be neglected. If its equilibrium position is taken as the (x, y) plane, then we are concerned with displacements $z(xy)$ perpendicular to this plane. Consider a small rectangular element $ABCD$ as in Fig. 11 of sides δx, δy. When this is vibrating the forces on it are (a)

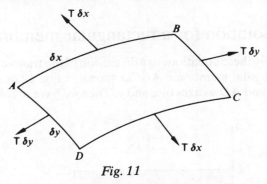

Fig. 11

two forces $\mathsf{T}\,\delta x$ perpendicular to AB and CD, and (b) two forces $\mathsf{T}\,\delta y$ perpendicular to AD and BC. These four forces act in the four tangent planes through the edges of the element. An argument precisely similar to that used in Chapter 2, **§17**, shows that the forces (a) give a resultant $\mathsf{T}\,\delta x.\dfrac{\partial^2 z}{\partial y^2}\,\delta y$ perpendicular to the plate. Similarly, the forces

(b) reduce to a force $\mathsf{T}\,\delta y.\dfrac{\partial^2 z}{\partial x^2}\,\delta x$. Let the mass of the plate be ρ per unit area; then, neglecting gravity, its equation of motion is

$$\mathsf{T}\frac{\partial^2 z}{\partial y^2}\,\delta x\,\delta y + \mathsf{T}\frac{\partial^2 z}{\partial x^2}\,\delta x\,\delta y = \rho\,\delta x\,\delta y\frac{\partial^2 z}{\partial t^2},$$

or

$$\mathsf{T}\left\{\frac{\partial^2 z}{\partial x^2}+\frac{\partial^2 z}{\partial y^2}\right\}=\rho\frac{\partial^2 z}{\partial t^2}.$$

This may be put in the standard form of the wave equation in two spatial dimensions

$$\frac{\partial^2 z}{\partial x^2}+\frac{\partial^2 z}{\partial y^2}=\frac{1}{c^2}\frac{\partial^2 z}{\partial t^2}, \tag{1}$$

where

$$c^2 = \mathsf{T}/\rho. \tag{2}$$

Thus we have reduced our problem to the solution of the standard wave equation and shown that the speed of waves along such membranes is $c = \sqrt{(\mathsf{T}/\rho)}$.

§33   Solution for a rectangular membrane

Let us apply these equations to a discussion of the transverse vibrations of a rectangular membrane *ABCD* shown in Fig. 12 of sides a and b. Take *AB* and *AD* as axes of x and y. Then we have to solve (1) subject

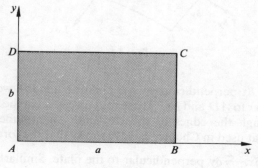

Fig. 12

to certain boundary conditions. These are that $z = 0$ at the boundary of the membrane, for all t. With our problem this means that $z = 0$ when $x = 0$, $x = a$, $y = 0$, $y = b$, independent of the time. The most suitable solution of the equation of wave motion is that of **§8**, equation (29). It is

$$z = \frac{\cos}{\sin} px \frac{\cos}{\sin} qy \frac{\cos}{\sin} rct, \qquad p^2 + q^2 = r^2.$$

If z is to vanish at $x = 0$, $y = 0$, we shall have to reject the cosines in the first two factors. Further, if z vanishes at $x = a$, then $\sin pa = 0$, so that $p = m\pi/a$, and similarly $q = n\pi/b$, m and n being positive integers. Thus the normal modes of vibration may be written

$$z = A \sin \frac{m\pi x}{a} \sin \frac{n\pi y}{b} \cos (rct + \varepsilon), \tag{3}$$

where

$$r^2 = (m^2/a^2 + n^2/b^2)\pi^2.$$

We may call this the (m, n) normal mode. Its frequency is $rc/2\pi$, or

$$\sqrt{\left\{ \left(\frac{m^2}{a^2} + \frac{n^2}{b^2} \right) \frac{\mathsf{T}}{4\rho} \right\}}. \tag{4}$$

The fundamental vibration is the $(1, 1)$ mode, for which the frequency is

$$\sqrt{\left\{ \left(\frac{1}{a^2} + \frac{1}{b^2} \right) \frac{\mathsf{T}}{4\rho} \right\}}.$$

The overtones (4) are not related in any simple numerical way to the fundamental and for this reason the sound of a vibrating plate, in which as a rule several modes are excited together, is much less musical to the ear than a string, where the harmonics are all simply related to the fundamental.

In the (m, n) mode of (3) there are nodal lines $x = 0$, a/m, $2a/m, \ldots, a$, and $y = 0$, b/n, $2b/n, \ldots, b$. On opposite sides of any nodal line the displacement has opposite sign. A few normal modes are shown in Fig. 13, in which the shaded parts are displaced oppositely to the unshaded.

The complete solution is the sum of any number of terms such as (3), with the constants chosen to give any assigned shape when $t = 0$. The method of choosing these constants is very similar to that of **§22**,

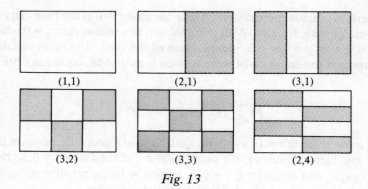

(1,1)   (2,1)   (3,1)

(3,2)   (3,3)   (2,4)

Fig. 13

except that there are now two variables x and y instead of one, and consequently we have double integrations corresponding to (27).

According to (4) the frequencies of vibration depend upon the two variables m and n. As a result it may happen that there are several different modes having the same frequency. Thus, for a square plate the (4, 7), (7, 4), (1, 8) and (8, 1) modes have the same frequency: and for a plate for which $a = 3b$, the (3, 3) and (9, 1) modes have the same frequency. When we have two or more modes with the same frequency, we call it a **degenerate** case. It is clear that any linear combination of these modes gives another vibration with the same frequency.

§34   Normal coordinates for a rectangular membrane

We can introduce normal coordinates as in the case of a vibrating string (cf. **§26**). According to (3) the full expression for z is

$$z = \sum_{m,n} A_{mn} \cos (rct + \varepsilon_r) \sin \frac{m\pi x}{a} \sin \frac{n\pi y}{b}. \tag{5}$$

We write this

$$z = \sum_{m,n} \phi_{mn} \sin \frac{m\pi x}{a} \sin \frac{n\pi y}{b}, \tag{6}$$

where ϕ_{mn} are the normal coordinates. The kinetic energy is

$$\int \int \tfrac{1}{2}\rho \left(\frac{\partial z}{\partial t}\right)^2 dx \, dy, \tag{7}$$

and this is easily shown to be

$$T = \sum_{m,n} \tfrac{1}{8}\rho ab\dot{\phi}_{mn}^2, \tag{8}$$

where a dot signifies differentiation with respect to time. The potential energy may be calculated in a manner similar to §14. Referring to Fig. 11 we see that in the displacement to the bent position, the two tensions $T\,\delta y$ have done work $T\,\delta y$. $(arc\,AB - \delta x)$. As in **§17**, this reduces to approximately $\tfrac{1}{2}T\left(\dfrac{\partial z}{\partial x}\right)^2 \delta x\,\delta y$. The other two tensions $T\,\delta x$ have done work $\tfrac{1}{2}T(\partial z/\partial y)^2\,\delta x\,\delta y$. The total potential energy is therefore

$$V = \tfrac{1}{2}T\iint\left\{\left(\frac{\partial z}{\partial x}\right)^2 + \left(\frac{\partial z}{\partial y}\right)^2\right\}\,\mathrm{d}x\,\mathrm{d}y. \tag{9}$$

In the case of the rectangular membrane this reduces to

$$V = \sum_{m,n} \tfrac{1}{8}\rho abc^2 r^2 \phi_{mn}^2. \tag{10}$$

It will be seen that T and V are both expressed in the form of Chapter 2, equation (38), typical of normal coordinates in mechanical problems.

§35   Circular membrane

With a circular membrane such as a drum of radius a, we have to use plane polar coordinates r, θ instead of Cartesians, and the solution of equation (1), apart from an arbitrary amplitude, is given in **§8**, equation (35a). It is

$$z = J_m(nr)\,\genfrac{}{}{0pt}{}{\cos}{\sin}\,m\theta\cos nct.$$

We have omitted the $Y_m(nr)$ term since this is not finite at $r = 0$. If we choose the origin of θ properly, this normal mode may be written

$$z = J_m(nr)\cos m\theta\cos nct. \tag{11}$$

If z is to be single-valued, m must be a positive integer. The boundary condition at $r = a$ is that for all values of θ and t, $J_m(na)\cos m\theta\cos nct$ equals zero. So that $J_m(na) = 0$. For any assigned value of m this equation has an infinite number of real roots, each one of which determines a corresponding value of n. These roots may be found from tables of Bessel functions. If we call them $n_{m,1}, n_{m,2} \ldots n_{m,k}, \ldots$, then the frequency of (11) is $nc/2\pi$, or $cn_{m,k}/2\pi$, and we may call it the (m, k) mode. The allowed values of m are $0, 1, 2, \ldots$ and of k are $1, 2, 3, \ldots$. There are nodal lines which consist of circles and radial lines. Figure 14 shows a few of these modes of vibration, shaded parts being displaced in an opposite direction to unshaded parts.

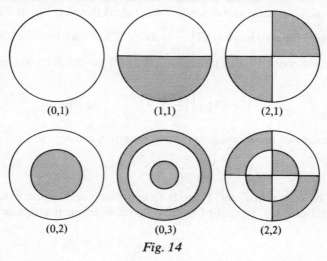

(0,1)   (1,1)   (2,1)

(0,2)   (0,3)   (2,2)

Fig. 14

The nodal lines obtained in Figs. 13 and 14 are known as Chladni's figures. A full solution of a vibrating membrane is obtained by superposing any number of these normal modes, and if nodal lines exist at all, they will not usually be of the simple patterns shown in these figures. As in the case of the rectangular membrane so also in the case of the circular membrane, the overtones bear no simple numerical relation to the fundamental frequency, and thus the sound of a drum is not very musical. A vibrating bell, however, is of very similar type, but it can be shown that some of the more important overtones bear a simple numerical relation to the fundamental; this would explain the pleasant sound of a well-constructed bell. But it is a little difficult to see why the ear so readily rejects some of the other overtones whose frequencies are not simply related to the fundamental. A possible explanation is that the mode of striking may be in some degree unfavourable to these discordant frequencies. In any case, we can easily understand why a bell whose shape differs slightly from the conventional, will usually sound unpleasant.

§36 Uniqueness of solutions

The uniqueness of the solution of a mixed initial and boundary value problem involving a vibrating plane membrane may be established for an arbitrary shape of membrane by means of a generalisation of the energy integral approach of §30. The task of showing this in the case of a rectangular plane membrane is left as an exercise for the reader (§37, Example 7). It should, however, be noticed that if only boundary data is given the solution will not be unique, for then any normal mode, or eigenfunction, will be a possible solution. Only when Cauchy data is given in addition, comprising the specification of z and $\partial z/\partial t$ over the membrane at time $t = 0$, is there sufficient information available to determine a unique solution.

The same is true, of course, of the vibrating string clamped at each end, where again the boundary conditions by themselves suffice only to determine the normal modes. The task of determining how to specify sufficient information in order that a solution of a partial differential equation should be unique is central to the study of such equations. Something of the important physical consequences of these matters should already be apparent from our study of waves on strings and membranes.

§37 Examples

1. Find two normal modes which are degenerate (§33) for a rectangular membrane of sides 6 and 3.

2. Obtain expressions for the kinetic and potential energies of a vibrating circular membrane. Perform the integrations over the θ-coordinate for the case of the normal mode

$$z = AJ_m(nr) \cos m\theta \cos mct.$$

3. A rectangular drum is $0.1\,\text{m} \times 0.2\,\text{m}$. It is stretched to a tension of $5000\,\text{Nm}^{-1}$, and its mass is $0.02\,\text{kg}$. What is the fundamental frequency?

4. A square membrane bounded by $x = 0$, a and $y = 0$, a is distorted into the shape

$$z = A \sin \frac{2\pi x}{a} \sin \frac{3\pi y}{a}$$

and then released. What is the resulting motion?

5. A rectangular membrane of sides a and b is stretched unevenly so that the tension in the x direction is T_1 and in the y direction is T_2. Show that the equation of motion is

$$T_1 \frac{\partial^2 z}{\partial x^2} + T_2 \frac{\partial^2 z}{\partial y^2} = \rho \frac{\partial^2 z}{\partial t^2}.$$

Show that this can be brought into the standard form by changing to new variables $x/\sqrt{T_1}$, $y/\sqrt{T_2}$, and hence find the normal modes.

6. Show that the number of normal modes for the rectangular membrane of §33 whose frequency is less than N is approximately equal to the area of a quadrant of the ellipse

$$\frac{x^2}{a^2} + \frac{y^2}{b^2} = \frac{4\rho}{T} N^2.$$

Hence show that the number is roughly $\pi \rho a b N^2 / T$.

7. Consider the energy

$$E(t) = \iint_D \left\{ \frac{1}{2}\rho \left(\frac{\partial z}{\partial t}\right)^2 + \frac{1}{2}F \left[\left(\frac{\partial z}{\partial x}\right)^2 + \left(\frac{\partial z}{\partial y}\right)^2\right] \right\} \, dx \, dy$$

associated with a vibrating rectangular membrane stretched over the region D with boundary ∂D comprising the rectangle $0 \leq x \leq a$, $0 \leq y \leq b$. Its motion is governed by the wave equation

$$\frac{\partial^2 z}{\partial t^2} = c^2 \left(\frac{\partial^2 z}{\partial x^2} + \frac{\partial^2 z}{\partial y^2}\right),$$

and it is clamped along ∂D so that on this boundary it satisfies the conditions

$$z = 0 \quad \text{on} \quad x = 0, \qquad x = a, \qquad y = 0 \quad \text{and} \quad y = b.$$

It also satisfies the initial conditions

$$z(x, y, 0) = h(x, y) \quad \text{and} \quad \frac{\partial z}{\partial t}(x, y, 0) = k(x, y).$$

By consideration of dE/dt show, by means of Green's theorem, that the subsequent motion is unique.

4

Longitudinal waves in bars and springs

§38 Differential equation for waves along a bar

The vibrations which we have so far considered have all been transverse, so that the displacement has been perpendicular to the direction of wave propagation. We must now consider **longitudinal waves**, in which the displacement is in the same direction as the wave. Suppose that AB in Fig. 15 is a bar of uniform section and mass ρ per unit length. The passage of a longitudinal wave along the bar will be represented by the vibrations of each element along the rod, instead of perpendicular to it. Consider a small element PQ of length δx, such that $AP = x$, and let us calculate the forces on this element, and hence its equation of motion, when it is displaced to a new position $P'Q'$. If the displacement of P to P' is ξ, then that of Q to Q' will be $\xi + \delta\xi$, so that $P'Q' = \delta x + \delta\xi$. We must first evaluate the tension at P'. We can do this by imagining δx to shrink to zero. Then the infinitesimally small element around P' will be in a state of tension T where, by Hooke's Law,

$$T_{P'} = \lambda \frac{\text{extension}}{\text{orig. length}}$$

$$= \lambda \operatorname*{Lim}_{\delta x \to 0} \frac{\delta x + \delta\xi - \delta x}{\delta x}$$

$$= \lambda \frac{\partial\xi}{\partial x}. \tag{1}$$

Returning to the element $P'Q'$, we see that its mass is the same as that of PQ, that is $\rho\,\delta x$, and its acceleration is $\dfrac{\partial^2\xi}{\partial t^2}$. Therefore

$$\rho\,\delta x \frac{\partial^2\xi}{\partial t^2} = T_{Q'} - T_{P'}$$

$$= \frac{\partial T}{\partial x}\delta x = \lambda \frac{\partial^2\xi}{\partial x^2}\delta x, \quad \text{by (1).}$$

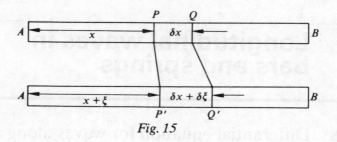

Fig. 15

Thus the equation of motion for these longitudinal waves reduces to the usual wave equation

$$\frac{\partial^2 \xi}{\partial x^2} = \frac{1}{c^2} \frac{\partial^2 \xi}{\partial t^2}, \quad \text{where } c^2 = \lambda/\rho. \tag{2}$$

The speed of waves along a rod is therefore $\sqrt{(\lambda/\rho)}$, a result similar in form to that for the speed of transverse oscillations of a string.

The full solution of (2) is soon found if we know the boundary conditions.

(i) At a free end the tension must vanish, and thus, from (1), $\partial \xi/\partial x = 0$, but the displacement will not, in general, vanish as well.

(ii) At a fixed end the displacement ξ must vanish, but the tension will not, in general, vanish also.

§39   Free vibrations of a finite bar

If we are interested in the free vibrations of a bar of length l, we shall use stationary type solutions of (2) as in §8, equation (27). Thus

$$\xi = (a \cos px + b \sin px) \cos \{cpt + \varepsilon\}.$$

If we take the origin at one end, then by (i) $\partial \xi/\partial x$ has to vanish at $x = 0$ and $x = l$. This means that $b = 0$, and $\sin pl = 0$, so that $pl = n\pi$, where $n = 1, 2, \ldots$. The free modes are therefore described by the functions

$$\xi_n = a_n \cos \frac{n\pi x}{l} \cos \left\{ \frac{n\pi ct}{l} + \varepsilon_n \right\}. \tag{3}$$

This normal mode has frequency $nc/2l$, so that the fundamental frequency is $c/2l$, and the harmonics are simply related to it. There are nodes in (3) at the points $x = l/2n, 3l/2n, 5l/2n, \ldots (2n-1)l/2n$; and there are antinodes (§6) at $x = 0, 2l/2n, 4l/2n \ldots l$. From (1) it follows that these positions are interchanged for the tension, nodes of motion being antinodes of tension and conversely. We shall meet this phenomenon again in Chapter 6.

§40   Vibrations of a clamped bar

The case of a rod rigidly clamped at its two ends is similarly solved. The boundary conditions are now that $\xi = 0$ at $x = 0$, and at $x = l$. The appropriate solution of (2) is thus

$$\xi_n = a_n \sin \frac{n\pi x}{l} \cos \left\{ \frac{n\pi ct}{l} + \varepsilon_n \right\}. \tag{4}$$

This solution has the same form as that found in Chapter 2, §**23**, for the transverse vibrations of a string.

§41   Normal coordinates

We may introduce normal coordinates for these vibrations, just as in §§**26** and **34**. Taking, for example, the case of §**40**, we should write

$$\xi = \sum_{n=1}^{\infty} \phi_n \sin \frac{n\pi x}{l}, \tag{5}$$

where

$$\phi_n = a_n \cos \left\{ \frac{n\pi ct}{l} + \varepsilon_n \right\}.$$

The kinetic energy of the element PQ is $\frac{1}{2}\rho\,\delta x \,.\, \dot{\xi}^2$, so that the total kinetic energy is

$$\int_0^l \tfrac{1}{2}\rho\dot{\xi}^2 \,\mathrm{d}x = \sum_n \tfrac{1}{4}\rho l\dot{\phi}_n^2. \tag{6}$$

The potential energy stored up in $P'Q'$ is approximately equal to one-half of the tension multiplied by the increase in length; or $\frac{1}{2}\lambda \dfrac{\partial \xi}{\partial x}\delta x$. Thus the total potential energy is

$$\int_0^l \frac{1}{2}\lambda\left(\frac{\partial \xi}{\partial x}\right)^2 \mathrm{d}x = \sum_n \frac{1}{4}\frac{\pi^2 n^2 c^2 \rho}{l}\phi_n^2. \tag{7}$$

§42 Case of a bar in a state of tension

The results of §§39, 40 for longitudinal vibrations of a bar need slight revision if the bar is initially in a state of tension. We shall discuss the vibrations of a bar of natural length l_0 stretched to a length l, so that its equilibrium tension T_0 is

$$T_0 = \lambda\left(\frac{l - l_0}{l_0}\right). \tag{8}$$

Referring to Fig. 15, we see that δx now represents the length of $P'Q'$ when in the stretched, non-vibrating state. The completely unstretched length is therefore not δx but $(l_0/l)\,\delta x$, so that tension at P' is not given by (1), but by the modified relation

$$T_{P'} = \lambda \lim_{\delta x \to 0} \frac{\delta x + \delta \xi - (l_0/l)\,\delta x}{(l_0/l)\,\delta x}$$

$$= T_0 + \frac{\lambda l}{l_0}\frac{\partial \xi}{\partial x}, \quad \text{using (8).} \tag{9}$$

The mass of PQ is $\rho_0(l_0/l)\,\delta x$ where ρ_0 refers to the unstretched bar, so the equation of motion is

$$\rho_0(l_0/l)\,\delta x\frac{\partial^2 \xi}{\partial t^2} = T_{Q'} - T_{P'} = \frac{\partial T}{\partial x}\delta x$$

$$= \frac{\lambda l}{l_0}\frac{\partial^2 \xi}{\partial x^2} \quad \text{from (9).}$$

We have again arrived at the standard wave equation

$$\frac{\partial^2 \xi}{\partial x^2} = \frac{1}{c^2}\frac{\partial^2 \xi}{\partial t^2}, \quad \text{where } c^2 = \lambda l^2/\rho_0 l_0^2. \tag{10}$$

It follows that $c = (l/l_0)c_0$, where c_0 is the velocity under no permanent tension. Appropriate solutions of (10) are soon seen to be

$$\xi_n = a_n \sin\frac{n\pi x}{l}\cos\left\{\frac{n\pi ct}{l} + \varepsilon_n\right\}, \quad \text{for } n = 1, 2 \ldots. \tag{11}$$

The fundamental frequency is $c/2l$, which, from (10), can be written $c_0/2l_0$. Thus with a given bar, the frequency is independent of the amount of stretching.

The normal mode (11) has nodes where $x = 0, l/n, 2l/n, \ldots, l$. A complete solution of (10) is obtained by superposition of separate solutions of type (11).

§43   Vibrations of a loaded spring

We now offer a discussion of the vibrations of a spring suspended from its top end and carrying a load M at its bottom end. When we neglect the mass of the spring it is easy to show that the lower mass M in Fig. 16

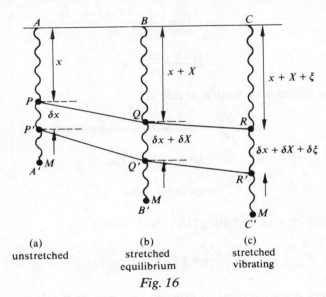

(a)	(b)	(c)
unstretched	stretched	stretched
	equilibrium	vibrating

Fig. 16

executes *simple harmonic motion* in a vertical line. Let us, however, consider the possible vibrations when we allow for the mass m of the spring. Put $m = \rho l$, where ρ is the unstretched mass per unit and l is the unstretched length. We may consider the spring in three stages. In stage (a) we have the unstretched spring of length l. The element PP' of length δx is at a distance x from the top point A. In stage (b) we have the equilibrium position when the spring is stretched due to its own weight and the load at the bottom. The element PP' is now displaced to QQ'. P is displaced a distance X downwards and P' a distance $X + \delta X$. Lastly, in stage (c) we suppose that the spring is vibrating and the element QQ' is displaced to RR'. The displacements of Q and Q' from their equilibrium positions are ξ and $\xi + \delta\xi$. The new length RR' is therefore $\delta x + \delta X + \delta\xi$. The mass of the element is the same as the mass of PP', that is, $\rho\,\delta x$, and is, of course, the same in all three stages.

We are now in a position to determine the equation of motion of RR'. The forces acting on it are its weight downwards and the two tensions at R and R'. The tension T_R may be found from Hooke's Law, by assuming that δx is made infinitesimally small. Then, as in §42,

$$T_R = \lambda \frac{\text{extension}}{\text{orig. length}},$$

$$= \lambda \lim_{\delta x \to 0} \frac{(\delta x + \delta X + \delta \xi) - \delta x}{\delta x},$$

$$= \lambda \left(\frac{\partial X}{\partial x} + \frac{\partial \xi}{\partial x} \right). \tag{12}$$

So the equation of motion of RR' is

$$\rho \, \delta x \frac{\partial^2 \xi}{\partial t^2} = \text{resultant force downwards}$$

$$= g\rho \, \delta x + T_{R'} - T_R,$$

$$= g\rho \, \delta x + \frac{\partial T}{\partial x} \delta x.$$

Dividing by $\rho \, \delta x$ and using (12), this becomes

$$\frac{\partial^2 \xi}{\partial t^2} = g + \frac{\lambda}{\rho} \left(\frac{\partial^2 X}{\partial x^2} + \frac{\partial^2 \xi}{\partial x^2} \right).$$

This last equation must be satisfied by $\xi = 0$, since this is merely the position of equilibrium (b). So

$$0 = g + \frac{\lambda}{\rho} \frac{\partial^2 X}{\partial x^2}.$$

By subtraction we discover once more the standard wave equation

$$\frac{\partial^2 \xi}{\partial x^2} = \frac{1}{c^2} \frac{\partial^2 \xi}{\partial t^2}, \qquad c^2 = \frac{\lambda}{\rho} = \frac{\lambda l}{m}. \tag{13}$$

This result is very similar to that of §42. However, before we can solve (13) we must discuss the boundary conditions. There are two of these. Firstly, when $x = 0$, we must have $\xi = 0$ for all t. Secondly, when $x = l$, (i.e., the position of the mass M) we must satisfy the law of motion

$$M \left[\frac{\partial^2 \xi}{\partial t^2} \right]_{x=l} = Mg - [T]_{x=l}.$$

Using (12), this becomes

$$\left[\frac{\partial^2 \xi}{\partial t^2}\right]_{x=1} = g - \frac{\lambda}{M}\left[\frac{\partial X}{\partial x} + \frac{\partial \xi}{\partial x}\right]_{x=i}.$$

As before, this equation must be satisfied by $\xi = 0$, since this is just the equilibrium stage (b). Thus

$$0 = g - \frac{\lambda}{M}\left[\frac{\partial X}{\partial x}\right]_{x=l}.$$

So, by subtraction we obtain the final form of the second boundary condition

$$\left[\frac{\partial^2 \xi}{\partial t^2}\right]_{x=l} = -\frac{\lambda}{M}\left[\frac{\partial \xi}{\partial x}\right]_{x=i}. \tag{14}$$

The appropriate solution of (13) is

$$\xi = a \sin px \cos (pct + \varepsilon). \tag{15}$$

This gives $\xi = 0$ when $x = 0$, and therefore satisfies the first boundary condition. It also satisfies the other boundary condition (14) if

$$pl \tan pl = m/M. \tag{16}$$

By plotting the curves $y = \tan x$, $y = (m/M)/x$, we see that there are solutions of (16) giving values of pl in the ranges 0 to $\pi/2$, π to $3\pi/2, \ldots$. The solutions become progressively nearer to $n\pi$ as n increases.

We are generally interested in the fundamental, or lowest, frequency, since this represents the natural vibrations of M at the end of the spring. The harmonics represent standing waves in the spring itself, and may be excited by gently stroking the spring downwards when in stage (b). If m/M is small, the lowest root of (16) is small; writing $pl = z$, we may expand $\tan z$ and get

$$z(z + z^3/3 + \ldots) = m/M.$$

Approximately

$$z^2(1 + z^2/3) = m/M.$$

We may put z^2 in the term in brackets equal to the first order approximation $z^2 = m/M$, and then we find for the second order approximation

$$z^2 = \frac{m/M}{1 + m/3M}.$$

The period of the lowest frequency in (15) is $2\pi/pc$, or, $2\pi l/cz$. Using the fact that $c^2 = \lambda l/m$, this becomes

$$2\pi \sqrt{\frac{l(M + \frac{1}{3}m)}{\lambda}}.$$

If the mass of the spring m had been neglected we should have obtained the result $2\pi\sqrt{(lM/\lambda)}$. It thus appears that the effect of the mass of the spring is equivalent, in a close approximation, to adding a mass one-third as great to the bottom of the spring.

§44   Waves in an anharmonic lattice

We conclude this chapter by making a brief study of longitudinal vibrational waves that can occur in a long coupled chain of spring and mass systems. Each individual system in the chain will be taken to be of length L and to comprise N equal masses m attached in a line, one to the other, by weightless identical springs. When a large number of these systems are connected to make a long chain it will be sufficient for us to consider the motion in only one such system. This follows because if we apply suitable periodic initial conditions there will then be periodicity of behaviour with any translation of length L, in the sense that any two masses a distance L apart in the chain will exhibit the same motion at any given time.

A mathematical model of this type can be used to describe the atomic vibrations in crystalline solids which, because of their regular structure, give rise to extremely long periodic chains. Regular structures of this nature are known as **lattices**, and their study in terms of the discrete model just outlined is called **lattice dynamics**. When the spring coupling is assumed to obey Hooke's Law these are called **harmonic lattices**, but when the spring behaviour is nonlinear they are then said to form **anharmonic lattices**. Our purpose here will be to construct a *continuum approximation* to a simple anharmonic lattice. These occur, for example, in the study of certain materials that exhibit anomalous heat conduction properties, and for which a continuum description is found to be more appropriate than a discrete one.

The situation in the lattice under consideration is illustrated in Fig. 17 where, in the undisturbed state, the masses m are all at a distance $h = L/N$ apart. We shall let x, measured from point A as origin, be the general position along the chain, while y_j denotes the displacement of the jth mass from its equilibrium position. The nonlinear spring will be taken to be described by the simple quadratic law

$$T = \varkappa(\Delta + \alpha\,\Delta^2), \tag{17}$$

where Δ is the displacement from the equilibrium position and $\varkappa, \alpha$ are constants.

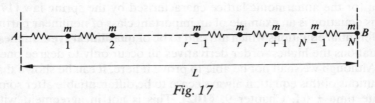

Fig. 17

If a dot is used to denote differentiation with respect to time, it is readily shown that the equation of motion of the jth mass in

$$m\ddot{y}_j = \varkappa(y_{j+1} - y_j) - \varkappa(y_j - y_{j-1}) + \varkappa\alpha[(y_{j+1} - y_j)^2 - (y_j - y_{j-1})^2], \tag{18}$$

which follows by considering the forces exerted on the jth mass by the two adjacent nonlinear springs after displacement. An initial value problem for this lattice then amounts to specifying $(y_j)_{t=0}$ and $(\dot{y}_j)_{t=0}$ for $j = 1, 2, \ldots, N$.

To reduce this discrete description to a continuum approximation we now assume that Taylor's theorem may be used to interpret $y_{j\pm1}$ in terms of partial derivatives of y_j with respect to x. The justification for this follows from the closeness of the masses and the smallness of the displacements y_j. We thus start from the relationships

$$y_{j+1} - y_j = \left[\, h\,\frac{\partial y}{\partial x} + \frac{h^2}{2!}\frac{\partial^2 y}{\partial x^2} + \frac{h^3}{3!}\frac{\partial^3 y}{\partial x^3} + \frac{h^4}{4!}\frac{\partial^4 y}{\partial x^4} + O(h^5)\right]$$

and

$$y_j - y_{j-1} = \left[h\frac{\partial y}{\partial x} - \frac{h^2}{2!}\frac{\partial^2 y}{\partial x^2} + \frac{h^3}{3!}\frac{\partial^3 y}{\partial x^3} - \frac{h^4}{4!}\frac{\partial^4 y}{\partial x^4} + O(h^5)\right].$$

Employing these results in (18), and neglecting terms of order greater than $O(h^3)$, we arrive at the nonlinear partial differential equation

$$\frac{\partial^2 y}{\partial t^2} = c^2 \left\{ 1 + 2\alpha h \left(\frac{\partial y}{\partial x} \right)^2 \right\} \frac{\partial^2 y}{\partial x^2},$$ (19)

where $c^2 = \varkappa h^2 / m$. The initial conditions for the lattice are then replaced by the specification of $(y)_{t=0}$ and $(\partial y / \partial x)_{t=0}$ on the initial line $0 < x \le L$, with periodicity conditions at the ends of the interval.

When the spring law is linear, so that $\alpha = 0$, the continuum approximation (19) reduces to the standard wave equation in which $c^2 = \varkappa h^2 / m$. Equation (19) represents the simplest continuum approximation for the anharmonic lattice characterised by the spring law (17). This equation is an example of an important class of nonlinear partial differential equations which are said to be of **quasilinear type**. In these equations the highest order derivatives all occur only to degree one.

Although we shall not be able to prove it here, it can be shown that solutions of this equation always cease to be differentiable after some finite time τ (cf. Chapter 9, **§102**). This is not in agreement with experimental observations of anharmonic lattices so that equation (19) cannot be regarded as providing a satisfactory description of the lattice for times approaching τ. To improve the approximation it is necessary to retain more terms when the Taylor series are substituted into the equation of motion (18).

If, instead of neglecting terms of order greater than $O(h^3)$, we retain one more term and neglect terms of order greater than $O(h^4)$, we obtain in place of (19) the equation

$$\frac{\partial^2 y}{\partial t^2} = c^2 \left[\frac{\partial^2 y}{\partial x^2} + 2\alpha h \left(\frac{\partial y}{\partial x} \right)^2 \frac{\partial^2 y}{\partial x^2} + \frac{h^2}{12} \frac{\partial^4 y}{\partial x^4} \right],$$ (20)

where again $c^2 = \varkappa h^2 / m$.

This is a fourth-order equation of degree one in its highest derivative so that this, too, is a quasilinear equation. Its form may be simplified by means of a change of variable to

$$\frac{\partial u}{\partial \tau} + u \frac{\partial u}{\partial \xi} + \mu \frac{\partial^3 u}{\partial \xi^3} = 0,$$ (21)

where

$$\xi = x - ct, \quad \tau = c\alpha h^2 t, \quad \mu = h/24\alpha \quad \text{and} \quad u = \partial y / \partial \xi.$$

This is the celebrated **Korteweg–de Vries** (KdV) **equation** first derived by Korteweg and de Vries in 1895 in connection with long water waves in shallow channels.

As this equation cannot be studied here, it must suffice for us to remark that its solutions do not suffer from the differentiability difficulties associated with solutions of equation (20), and it can be used as a continuum approximation to an anharmonic lattice.

The anharmonic lattice has thus provided us with an example of a wave type physical situation for which a linear wave equation is an unacceptable mathematical model. Furthermore, although we were unable to demonstrate it in this account, only the second nonlinear approximation can be regarded as being in any way satisfactory. This situation contrasts sharply with the way we were able to approximate the equation of the string in Chapter 2, §17 by the standard wave equation.

§45 Examples

1. Find the velocity of longitudinal waves along a bar whose mass is $0 \cdot 225 \text{ kg m}^{-1}$ and for which the modulus is 9×10^5 N.

2. Consider a bar in which the cross-section at position x along its length is a function $S(x)$ of x only. Reformulate the equation of motion in §38 for an element of the bar to show that it takes the form

$$\frac{\partial T}{\partial x} \delta x = S \rho_0 \, \delta x \frac{\partial^2 \xi}{\partial x^2},$$

where now ρ_0 is the density per unit volume of the material of the bar. Show also that Hooke's Law implies that the tensile force T is given by

$$T = S(x) Y \frac{\partial \xi}{\partial x},$$

where Y is a constant for the material of the bar. It is known as **Young's modulus**. Hence deduce that when the cross-section of the bar varies with x, the displacement ξ must satisfy the equation

$$\frac{\partial^2 \xi}{\partial x^2} + \frac{1}{S}\left(\frac{\mathrm{d}S}{\mathrm{d}x}\right)\frac{\partial \xi}{\partial x} = \frac{1}{c^2}\frac{\partial^2 \xi}{\partial t^2},$$

where $c^2 = Y/\rho_0$.

3. Waves are transmitted from an ultrasonic generator to their point of application along a bar whose cross-section S obeys the equation

$$S(x) = S_0 \, \mathrm{e}^{-2ax} \quad \text{with } a > 0.$$

Obtain a solution ξ to the equation derived in the previous example in the form $\xi(x, t) = h(x) \, e^{-int}$ and use it to show that when $n^2 > a^2 c^2$ the amplitude of the waves transmitted along the bar grows at an exponential rate.

4. Two semi-infinite bars are joined to form an infinite rod. Their moduli are λ_1 and λ_2 and the densities per unit length are ρ_1 and ρ_2. Investigate the reflection coefficient (see §20) and the phase change on reflection, when harmonic waves in the first medium meet the join of the bars.

5. Investigate the normal modes of a bar rigidly fastened at one end and free to move longitudinally at the other.

6. A uniform bar of length l is hanging freely from one end. Show that the frequencies of the normal longitudinal vibrations are $(n + \tfrac{1}{2}) \, c/2l$, where c is the velocity of longitudinal waves in the bar.

7. The modulus of a spring is $7 \cdot 2 \times 10^{-2}$ N. Its mass is $0 \cdot 01$ kg and its unstretched length is $0 \cdot 012$ m. A mass $0 \cdot 04$ kg is hanging on the lowest point, and the top point is fixed. Calculate to an accuracy of one per cent the periods of the lowest two vibrations.

8. Investigate the vertical vibrations of a spring of unstretched length $2l$ and mass $2m$, supported at its top end and carrying loads M at the mid-point and the bottom.

Waves in liquids

§46 Summary of hydrodynamical formulae

In this chapter we shall discuss wave motion in liquids. We shall assume that the liquid is incompressible, with constant density ρ. This condition is very nearly satisfied by most liquids, and the case of a compressible fluid is dealt with in Chapter 6. We shall further assume that the motion is irrotational. This is equivalent to neglecting viscosity and assuming that all the motions have started from rest due to the influence of natural forces such as wind, gravity, or pressure of certain boundaries. If the motion is irrotational, we may assume the existence of a velocity potential ϕ if we desire it. It will be convenient to summarise the formulae which we shall need in this work.

(i) If the vector **u** with components (u, v, w) represents the velocity of any part of the fluid, then from the definition of ϕ

$$\mathbf{u} = -\nabla\phi \equiv -\text{grad } \phi, \tag{1}$$

so that in particular $u = -\partial\phi/\partial x$, $v = -\partial\phi/\partial y$, $w = -\partial\phi/\partial z$.

(ii) On a fixed boundary the velocity has no normal component, and hence if $\partial/\partial\nu$ denotes differentiation along the normal,

$$\partial\phi/\partial\nu = 0. \tag{2}$$

(iii) Since no liquid will be supposed to be created or annihilated, the equation of continuity must express the conservation of mass; it is

$$\nabla \cdot \mathbf{u} \equiv \frac{\partial u}{\partial x} + \frac{\partial v}{\partial y} + \frac{\partial w}{\partial z} = 0. \tag{3}$$

Combining (1) and (3), we obtain Laplace's equation

$$\nabla^2\phi \equiv \frac{\partial^2\phi}{\partial x^2} + \frac{\partial^2\phi}{\partial y^2} + \frac{\partial^2\phi}{\partial z^2} = 0. \tag{4}$$

(iv) If $H(x, y, z, t)$ is any property of a particle of the fluid, such as its velocity, pressure or density, then $\partial H/\partial t$ is the variation of H at a *particular point in space*, and

$$\frac{DH}{Dt}$$

is the variation of H *when we keep to the same particle of fluid*. This quantity is known as the **total derivative**, or the **material derivative following the fluid**, and it can be shown that

$$\frac{DH}{Dt} = \frac{\partial H}{\partial t} + \mathbf{u} \cdot \nabla H$$

or,

$$\frac{DH}{Dt} = \frac{\partial H}{\partial t} + u \frac{\partial H}{\partial x} + v \frac{\partial H}{\partial y} + w \frac{\partial H}{\partial z}. \qquad (5)$$

(v) If the external forces acting on unit mass of liquid can be represented by a vector $\mathbf{F}$, then the equation of motion of the liquid may be expressed in vector form

$$\frac{D\mathbf{u}}{Dt} = \mathbf{F} - \frac{1}{\rho} \nabla p.$$

In Cartesian form this is

$$\frac{\partial u}{\partial t} + u \frac{\partial u}{\partial x} + v \frac{\partial u}{\partial y} + w \frac{\partial u}{\partial z} = F_x - \frac{1}{\rho} \frac{\partial p}{\partial x}, \qquad (6)$$

with two similar equations for v and w.

(vi) An important integral of the equations of motion can be found in cases where the external force $\mathbf{F}$ has a potential V, so that $\mathbf{F} = -\nabla V$. The integral in question is known as **Bernoulli's Equation:**

$$\frac{p}{\rho} + \frac{1}{2}\mathbf{u}^2 + V - \frac{\partial \phi}{\partial t} = C, \qquad (7)$$

where C is an arbitrary function of the time. Now according to (1), addition of a function of t to ϕ does not affect the velocity distribution given by ϕ; it is often convenient, therefore, to absorb C into the term $\frac{\partial \phi}{\partial t}$ and (7) can then be written

$$\frac{p}{\rho} + \frac{1}{2}\mathbf{u}^2 + V - \frac{\partial \phi}{\partial t} = \text{const.} \qquad (8)$$

A particular illustration of (8) which we shall require later occurs at the surface of water waves; here the pressure must equal the atmospheric pressure and is hence constant. Thus at the surface of the waves (sometimes called the **free surface**)

$$\frac{1}{2}\mathbf{u}^2 + V - \frac{\partial \phi}{\partial t} = \text{constant}. \tag{9}$$

§47 Tidal waves and surface waves

We may divide the types of wave motion in liquids into two groups; the one group has been called **tidal waves**, and arises when the wavelength of the oscillations is much greater than the depth of the liquid. Another name for these waves is **long waves in shallow water**. With waves of this type the vertical acceleration of the liquid is neglected in comparison with the horizontal acceleration, and we shall be able to show that liquid originally in a vertical plane remains in a vertical plane throughout the vibrations; thus each vertical plane of liquid moves as a whole. The second group may be called **surface waves**, and in these the disturbance does not extend far below the surface. The vertical acceleration is no longer negligible and the wavelength is much less than the depth of the liquid. To this group belong most wind waves and surface tension waves. We shall consider the two types separately, though it will be recognised that tidal waves represent an approximation and the results for these waves may often be obtained from the formulae of surface waves by introducing certain restrictions.

§48 Tidal waves, general conditions

We shall deal with *tidal waves* first. Here we assume that the vertical accelerations may be neglected. One important result follows immediately. If we draw the z axis vertically upwards (as we shall continue to do throughout this chapter), then the equation of motion in the z direction as given by (6), is

$$\frac{Dw}{Dt} = -g - \frac{1}{\rho}\frac{\partial p}{\partial z}.$$

We are to neglect $\dfrac{Dw}{Dt}$ and thus

$$\frac{\partial p}{\partial z} = -g\rho, \quad \text{so } p = -g\rho z + \text{constant}.$$

Let us take our (x, y) plane in the undisturbed free surface, and write $\zeta(x, y, t)$ for the elevation of the water above the point $(x, y, 0)$. Then, if the atmospheric pressure is p_0, we must have $p = p_0$ when $z = \zeta$. So the equation for the pressure becomes

$$p = p_0 + g\rho(\zeta - z). \tag{10}$$

We can put this value of p into the two equations of horizontal motion, and we obtain

$$\frac{Du}{Dt} = -g\frac{\partial\zeta}{\partial x}, \qquad \frac{Dv}{Dt} = -g\frac{\partial\zeta}{\partial y}. \tag{11}$$

The right-hand sides of these equations are independent of z, and we deduce therefore that in this type of motion the horizontal acceleration is the same at all depths. Consequently, as we stated earlier without proof, in shallow water the velocity does not vary with the depth, and the liquid moves as a whole, in such a way that particles originally in a vertical plane, remain so, although this vertical plane may move as a whole.

§49   Tidal waves in a straight channel

Let us now apply the results of the last section to discuss tidal waves along a straight horizontal channel whose depth is constant, but whose cross-section A varies from place to place. We shall suppose that the waves move in the x direction only (extension to two dimensions will come later). Consider the liquid in a small volume as in Fig. 18 bounded by the vertical planes $x, x + \delta x$ at P and Q. The liquid in the vertical plane through P is all moving with the same horizontal velocity $u(x)$ independent of the depth. We can suppose that A varies sufficiently slowly for us to neglect motion in the y direction. We have two equations with which to obtain the details of the motion. The first is (11) and may be written

$$\frac{\partial u}{\partial t} + u\frac{\partial u}{\partial x} + w\frac{\partial u}{\partial z} = -g\frac{\partial\zeta}{\partial x}.$$

Since u is independent of z, we have $\partial u/\partial z = 0$. Further, since we shall suppose that the velocity of any element of fluid is small, we may neglect $u(\partial u/\partial x)$ which is of the second order, and rewrite this

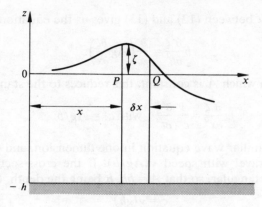

Fig. 18

equation

$$\frac{\partial u}{\partial t} = -g \frac{\partial \zeta}{\partial x}. \tag{12}$$

The second equation is the *equation of continuity*. Equation (3) is not convenient for this problem, but a suitable equation can be found by considering the volume of liquid between the planes at P and Q, in Fig. 18. Let $b(x)$ be the breadth of the water surface at P. Then the area of the plane P which is covered with water is $[A + b\zeta]_P$; therefore the amount of liquid flowing into the volume per unit time is $[(A + b\zeta)u]_P$. Similarly, the amount flowing out per unit time at Q is $[(A + b\zeta)u]_Q$. The difference between these is compensated by the rate at which the level is rising inside the volume, and thus

$$[(A + b\zeta)u)]_P - [(A + b\zeta)u]_Q = b \, \delta x \frac{\partial \zeta}{\partial t}.$$

Therefore

$$-\frac{\partial}{\partial x}\{(A + b\zeta)u\} = b \frac{\partial \zeta}{\partial t}.$$

Since $b\zeta u$ is of the second order of small quantities, we may neglect this term and the equation of continuity becomes

$$-\frac{\partial}{\partial x}(Au) = b \frac{\partial \zeta}{\partial t}. \tag{13}$$

Eliminating u between (12) and (13) gives us the equation

$$b\frac{\partial^2 \zeta}{\partial t^2} = \frac{\partial}{\partial x}\left(Ag\frac{\partial \zeta}{\partial x}\right). \tag{14}$$

In the case in which A is constant, this reduces to the standard form

$$\frac{\partial^2 \zeta}{\partial x^2} = \frac{1}{c^2}\frac{\partial^2 \zeta}{\partial t^2} \quad \text{with } c^2 = Ag/b. \tag{15}$$

This is the familiar wave equation in one dimension, and we deduce that waves travel with speed $\sqrt{(Ag/b)}$. If the cross-section of the channel is rectangular, so that $A = bh$, h being the depth, then

$$c = \sqrt{(gh)}. \tag{16}$$

With an unlimited channel, there are no boundary conditions involving x, and to our degree of approximation waves with any profile will travel in either direction. With a limited channel, there will be boundary conditions. Thus, if the ends are vertical, $u = 0$ at each of them.

We may apply this to a rectangular basin of length l, whose two ends are at $x = 0, l$. Possible solutions of (15) are given in §8, equation (27). They are

$$\zeta = (a \cos px + \beta \sin px) \cos (cpt + \varepsilon).$$

Then, using (13) and also the fact that $A = bh$, we find

$$\frac{\partial u}{\partial x} = \frac{cp}{h}(a \cos px + \beta \sin px) \sin (cpt + \varepsilon).$$

and so

$$u = \frac{c}{h}(a \sin px - \beta \cos px) \sin (cpt + \varepsilon).$$

The boundary conditions $u = 0$ at $x = 0, l$, imply that $\beta = 0$, and $\sin pl = 0$. So

$$\zeta_r = a_r \cos \frac{r\pi x}{l} \cos \left\{\frac{r\pi ct}{l} + \varepsilon_r\right\}, \quad \text{for } r = 1, 2, 3, \dots \tag{17}$$

and

$$u_r = \frac{a_r c}{h} \sin \frac{r\pi x}{l} \sin \left\{\frac{r\pi ct}{l} + \varepsilon_r\right\}. \tag{18}$$

It will be noticed that nodes of u_r and ζ_r do not occur at the same points.

The vertical velocity may be found from the general form of the equation of continuity (3). Applied to our case, this is

$$\frac{\partial u}{\partial x} + \frac{\partial w}{\partial z} = 0.$$

Now u is independent of z and $w = 0$ on the bottom of the liquid where $z = -h$. Consequently, on integrating we find

$$w_r = -(z+h)\frac{\partial u}{\partial x} = \frac{-\pi r a_r c}{lh}(z+h)\cos\frac{r\pi x}{l}\sin\left\{\frac{r\pi ct}{l}+\varepsilon_r\right\}. \quad (19)$$

We may use this last equation to deduce under what conditions our original assumption that the vertical acceleration could be neglected, is valid. For similarly to (12), the vertical acceleration Dw_r/Dt is effectively $\partial w_r/\partial t$, giving

$$-\frac{\pi^2 r^2 c^2 a_r}{l^2 h}(z+h)\cos\frac{r\pi x}{l}\cos\left\{\frac{r\pi ct}{l}+\varepsilon_r\right\}.$$

The maximum value of this is $\pi^2 r^2 c^2 a_r/l^2$, and may be compared with the maximum horizontal acceleration $\pi r c^2 a_r/lh$. The ratio of the two for an arbitrary rth mode is $r\pi h/l$; that is $2\pi h/\lambda$, since, from (17) $\lambda = 2l/r$. We have therefore confirmed the condition which we stated as typical of these long waves, namely that the vertical acceleration may be neglected if the wavelength is much greater than the depth of water.

§50   Tidal waves on lakes and tanks

We shall now remove the restriction imposed in the last section to waves in one dimension. Let us use the same axes as before and consider the rate of flow of liquid into a vertical prism bounded by the planes $x, x+\delta x, y, y+\delta y$. In Fig. 19, *ABCD* is the undistributed surface, *EFGH* is the bottom of the liquid, and *PQRS* is the moving surface at height $\zeta(x, y)$ above *ABCD*. The rate of flow into the prism across the face *PEHS* is $[u(h+\zeta)\,\delta y]_x$, and the rate of flow out across *RQFG* is $[u(h+\zeta)\,\delta y]_{x+\delta x}$. The net result from these two planes is a gain

$$-\frac{\partial}{\partial x}\{u(h+\zeta)\,\delta x\,\delta y.$$

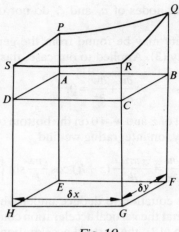

Fig. 19

Similarly, from the other two vertical planes there is a gain

$$-\frac{\partial}{\partial y}\{v(h+\zeta)\}\,\delta x\,\delta y.$$

The total gain is balanced by the rising of the level inside the prism, and thus

$$-\frac{\partial}{\partial x}\{u(h+\zeta)\}\,\delta x\,\delta y - \frac{\partial}{\partial y}\{v(h+\zeta)\}\,\delta x\,\delta y = \frac{\partial\zeta}{\partial t}\,\delta x\,\delta y.$$

As in §49 we may neglect terms such as $u\zeta$ and $v\zeta$, and thus write the above equation of continuity in the form

$$\frac{\partial(hu)}{\partial x} + \frac{\partial(hv)}{\partial y} = -\frac{\partial\zeta}{\partial t}. \tag{20}$$

We have to combine this equation with the two equations of motion (11), which yield, after neglecting square terms in the velocities,

$$\frac{\partial u}{\partial t} = -g\frac{\partial\zeta}{\partial x}, \qquad \frac{\partial v}{\partial t} = -g\frac{\partial\zeta}{\partial y}. \tag{21}$$

Eliminating u and v gives us the standard equation

$$\frac{\partial}{\partial x}\left(h\frac{\partial\zeta}{\partial x}\right) + \frac{\partial}{\partial y}\left(h\frac{\partial\zeta}{\partial y}\right) = \frac{1}{g}\frac{\partial^2\zeta}{\partial t^2}. \tag{22}$$

If h is constant (tank of constant depth) this becomes

$$\frac{\partial^2 \zeta}{\partial x^2} + \frac{\partial^2 \zeta}{\partial y^2} = \frac{1}{c^2} \frac{\partial^2 \zeta}{\partial t^2}, \qquad c^2 = gh. \tag{23}$$

This is the usual wave equation in two dimensions and shows that the velocity is $\sqrt{(gh)}$. If we are concerned with waves in one dimension, so that ζ is independent of y (as in §49) we put $\partial^2 \zeta / \partial y^2 = 0$ and retrieve (15).

We have therefore to solve the wave equation subject to the boundary conditions;

  (i) $w = 0$ at $z = -h$,

 (ii) $\dfrac{\partial \zeta}{\partial x} = 0$ at a boundary parallel to the y axis, and

  $\dfrac{\partial \zeta}{\partial y} = 0$ at a boundary parallel to the x axis,

(iii) $\dfrac{\partial \zeta}{\partial v} = 0$ at any fixed boundary, where $\dfrac{\partial}{\partial v}$ denotes differentiation along the normal to the boundary.

This latter condition, of which (ii) is a particular case, can be seen as follows. If $lx + my = 1$ is the fixed boundary, then the component of the velocity perpendicular to this line has to vanish. That is, $lu + mv = 0$. By differentiating partially with respect to t and using (21), the condition (iii) is obtained.

§51   Tidal waves on rectangular and circular tanks

We shall apply these formulae to two cases; first, a rectangular tank, and, second, a circular one, both of constant depth.

Rectangular tank.   Let the sides be $x = 0$, a and $y = 0$, b. Then a suitable solution of (23) satisfying all the boundary conditions (i) and (ii) would be

$$\zeta = A \cos \frac{p\pi x}{a} \cos \frac{q\pi y}{b} \cos (r\pi c t + \varepsilon), \tag{24}$$

where $p = 0, 1, 2 \ldots, q = 0, 1, 2, \ldots,$ and $r^2 = p^2/a^2 + q^2/b^2$. This solution closely resembles that for a vibrating membrane in Chapter 3,

§33, and the nodal lines are of the same general type. The reader will recognise how closely the solution (24) resembles a "choppy sea."

Circular tank.   If the centre of the tank is taken as the origin and its radius is a, then the boundary condition (iii) reduces to $\partial \zeta / \partial r = 0$ at $r = a$. Suitable solutions of (23) in polar coordinates have been given in Chapter 1, equation (35a). We have

$$\zeta = A \cos m\theta J_m(nr) \cos(cnt + \varepsilon). \tag{25}$$

We have rejected the Y_m solution since it is infinite at $r = 0$, and we have chosen the zero of θ so that there is no term in $\sin m\theta$. This expression satisfies all the conditions except the boundary condition (iii) at $r = a$. This requires that $J'_m(na) = 0$. For a given value of m (which must be integral) this condition determines an infinite number of values of n, whose magnitudes may be found from tables of Bessel Functions. The nodal lines are concentric circles and radii from the origin, very similar to those in Fig. 14 for a vibrating membrane. The period of this motion is $2\pi / cn$.

§52   Paths of particles

It is possible to determine the actual paths of individual particles in many of these problems. Thus, referring to the rectangular tank of §49 the velocities u and w are given by (18) and (19). We see that

$$\frac{w}{u} = \frac{-\pi r(z + h)}{l} \cot \frac{r\pi x}{l}.$$

This quantity is independent of the time and thus any particle of the liquid executes simple harmonic motion along a line whose slope is given by the above value of w/u. For particles at a fixed depth, this direction changes from purely horizontal beneath the nodes to purely vertical beneath the antinodes.

§53   Method of reduction to a steady wave

We shall conclude our discussion of tidal waves by applying the method of reduction to a steady wave, already described in §29, to the case of waves in a channel of constant cross-section A and breadth of water line b. This is the problem of §49 with A constant. Let c be the velocity of propagation of a wave profile. Then superimpose a velocity $-c$ on the whole system, so that the wave profile becomes stationary and the liquid flows under it with mean velocity c. The actual velocity at any point will differ from c since the cross-sectional area of the liquid is not constant. This area is $A + b\zeta$, and varies with ζ. Let the velocity be $c + \theta$ at sections where the elevation is ζ. Since no liquid is piling up, the volume of liquid crossing any plane perpendicular to the direction of flow is constant, so that

$$(A + b\zeta)(c + \theta) = \text{constant} = Ac. \tag{26}$$

We have still to use the fact that the pressure at the free surface is always atmospheric. In Bernoulli's equation at the free surface (9) we may put $\partial\phi/\partial t = 0$ since the motion is now steady motion; also $V = g\zeta$ at the free surface. So, neglecting squares of the vertical velocity, this gives

$$\tfrac{1}{2}(c + \theta)^2 + g\zeta = \text{const.} = \tfrac{1}{2}c^2.$$

Eliminating θ between this equation and (26), we have

$$\frac{A^2 c^2}{(A + b\zeta)^2} + 2g\zeta = c^2,$$

and so,

$$2g\zeta = c^2\left\{1 - \frac{A^2}{(A + b\zeta)^2}\right\} = bc^2\zeta\left\{\frac{2A + b\zeta}{(A + b\zeta)^2}\right\}.$$

Consequently we arrive at the result

$$c^2 = \frac{2g}{b}\frac{(A + b\zeta)^2}{2A + b\zeta}. \tag{27}$$

If ζ is small, so that we may neglect ζ compared with A/b, then this equation gives the same result as (16), namely, $c^2 = gA/b$. We can, however, deduce more than this simple result. For if $\zeta > 0$, the right-hand side of (27) is greater than gA/b, and if $\zeta < 0$, it is less than gA/b. Thus an elevation travels slightly faster than a depression and so it is impossible for a long wave of this type to be propagated without change of shape. Further, since the tops of waves travel faster than the troughs, we have an explanation of why waves break near the sea-shore when they reach shallow water.

§54   Surface waves, the velocity potential

We now consider *surface waves*, in which the restriction is removed that the wavelength is much greater than the depth. In these waves the disturbance is only appreciable over a finite depth of the liquid. We shall solve this problem by means of the velocity potential ϕ. We know ϕ must satisfy Laplace's equation (4) and at any fixed boundary $\partial\phi/\partial\nu = 0$, by (2). There are, however, two other conditions imposed on ϕ at the free surface. The first arises from Bernoulli's equation (9). If the velocity is so small that $\mathbf{u}^2$ may be neglected, and if the only external forces are the external pressure and gravity, we may put $\mathbf{u}^2 = 0$ and $V = g\zeta$ in this equation, which then becomes

$$\zeta = \frac{1}{g}\left[\frac{\partial\phi}{\partial t}\right]_{\text{free surface}}. \tag{28}$$

The second condition can be seen as follows. A particle of fluid originally on the free surface will remain so always. Now the equation of the free surface, where $z = \zeta(x, y, t)$ may be written

$$0 = f(x, y, z, t) = \zeta(x, y, t) - z.$$

Consequently, f is a function which is always zero for a particle on the free surface. We may therefore use (5) with H put equal to f, and we find

$$0 = \frac{Df}{Dt} = \frac{\partial\zeta}{\partial t} + u\frac{\partial\zeta}{\partial x} + v\frac{\partial\zeta}{\partial y} - w.$$

Now from (28)

$$\frac{\partial\zeta}{\partial x} = \frac{1}{g}\frac{\partial}{\partial t}\left(\frac{\partial\phi}{\partial x}\right) = -\frac{1}{g}\frac{\partial u}{\partial t}$$

on the surface.

Thus $\partial\zeta/\partial x$ is a small quantity of order of magnitude not greater than u; consequently $u(\partial\zeta/\partial x)$ and $v(\partial\zeta/\partial y)$ being of order of magnitude not greater than u^2, may be neglected. We are left with the new boundary condition

$$\frac{\partial\zeta}{\partial t} = w = -\frac{\partial\phi}{\partial z}. \tag{29}$$

Combining (28) and (29) we obtain an alternative relation

$$\frac{\partial^2\phi}{\partial t^2} + g\frac{\partial\phi}{\partial z} = 0. \tag{30}$$

We summarise the conditions satisfied by ϕ as follows:

(i) Laplace's equation $\nabla^2\phi = 0$ in the liquid, (4)

(ii) $\partial\phi/\partial\nu = 0$ on a fixed boundary, (2)

(iii) $\zeta = \dfrac{1}{g}\dfrac{\partial\phi}{\partial t}$ on the free surface, (28)

(iv) $\dfrac{\partial\zeta}{\partial t} = -\dfrac{\partial\phi}{\partial z}$ on the free surface, (29)

(v) $\dfrac{\partial^2\phi}{\partial t^2} + g\dfrac{\partial\phi}{\partial z} = 0$ on the free surface. (30)

Only two of the last three conditions are independent.

§55 Surface waves on a long rectangular tank

Let us apply these equations to the case of a liquid of depth h in an infinitely long rectangular tank, supposing that the motion takes place along the length of the tank, which we take as the x direction. The axes of x and y lie, as usual, in the undisturbed free surface. Condition (i) above gives an equation which may be solved by the method of separation of variables (see §7), and if we want our solution to represent a progressive wave with velocity c, a suitable form of the solution would be

$$\phi = (A\,e^{mz} + B\,e^{-mz})\cos m(x - ct).$$

A, B, m and c are to be determined from the other conditions (ii)–(v). At the bottom of the tank (ii) gives $\partial\phi/\partial z = 0$, or $A\,e^{-mh} - B\,e^{mh} = 0$. So $A\,e^{-mh} = B\,e^{mh} = \frac{1}{2}C$, say, and hence

$$\phi = C\cosh m(z + h)\cos m(x - ct). \tag{31}$$

Condition (v) applies at the free surface where, if the disturbance is not too large, we may put $z = 0$; after some reduction it becomes

$$c^2 = (g/m)\tanh mh.$$

Since $m = 2\pi/\lambda$, where l is the wavelength, we can write this

$$c^2 = \frac{g\lambda}{2\pi}\tanh\frac{2\pi h}{\lambda}. \tag{32}$$

Recalling from Chapter 1, §3, that $c = n/k$ and $\lambda = 1/k$, we recognise that (32) is merely the dispersion relation for this wave motion. It is this strong dispersive action in water waves that leads to the changing shape of waves as they progress.

Condition (iii) gives us the appropriate form of ζ; it is

$$\zeta = \frac{mcC}{g} \cosh mh \, \sin m(x - ct).$$

This expression becomes more convenient if we write a for the amplitude of ζ; when $a = (mcC/g) \cosh mh$. Then

$$\zeta = a \sin m(x - ct), \tag{33}$$

and

$$\phi = \frac{ga}{mc} \frac{\cosh m(z + h)}{\cosh mh} \cos m(x - ct). \tag{34}$$

If the water is very deep so that $\tanh (2\pi h/\lambda) = 1$, then (32) becomes $c^2 = g\lambda/2\pi$, and if it is very shallow so that $\tanh (2\pi h/\lambda) = 2\pi h/\lambda$, we retrieve the formula of **§49** for long waves in shallow water, giving $c^2 = gh$.

We have seen in Chapter 1 that stationary waves result from superposition of two opposite progressive harmonic waves. Thus we could have stationary waves analogous to (33) and (34) defined by

$$\zeta = a \sin mx \cos mct, \tag{35}$$

and

$$\phi = \frac{ga}{mc} \frac{\cosh m(z + h)}{\cosh mh} \sin mx \, \sin mct. \tag{36}$$

We could use these last two equations to discuss stationary waves in a rectangular tank of finite length.

§56   Surface waves in two dimensions

We shall now discuss surface waves in two dimensions, considering two cases in particular.

Rectangular tank.   With a rectangular tank bounded by the planes $x = 0, a$ and $y = 0, b$, it is easily verified that all the conditions of **§54** are satisfied by

$$\zeta = A \cos \frac{p\pi x}{a} \cos \frac{q\pi y}{b} \cos rct,$$

$$\phi = \frac{gA}{rc} \frac{\cosh r(z + h)}{\cosh rh} \cos \frac{p\pi x}{a} \cos \frac{q\pi y}{b} \sin rct,$$

where

$$p = 1, 2, \ldots ; q = 1, 2, \ldots ; r^2 = \pi^2(p^2/a^2 + q^2/b^2)$$

and

$$c^2 = (g/r) \tanh rh. \tag{37}$$

Circular tank. Suppose that the tank is of radius a and depth h. Then choosing the centre as origin and using cylindrical polar coordinates r, θ, z, Laplace's equation (cf. Chapter 1, §7) becomes

$$\frac{\partial^2 \phi}{\partial r^2} + \frac{1}{r} \frac{\partial \phi}{\partial r} + \frac{1}{r^2} \frac{\partial^2 \phi}{\partial \theta^2} + \frac{\partial^2 \phi}{\partial z^2} = 0. \tag{38}$$

A suitable solution can be found from Chapter 1, equation (35a), which gives us a solution of the similar equation

$$\frac{\partial^2 \phi}{\partial r^2} + \frac{1}{r} \frac{\partial \phi}{\partial r} + \frac{1}{r^2} \frac{\partial^2 \phi}{\partial \theta^2} - \frac{1}{c^2} \frac{\partial^2 \phi}{\partial t^2} = 0$$

in the form

$$\phi = \frac{J_m}{Y_m} (nr) \frac{\cos}{\sin} m\theta \frac{\cos}{\sin} nct.$$

In this equation let us make a change of variable, writing $ct = iz$, where $i^2 = -1$. We then get Laplace's equation (38) and its solutions are therefore

$$\phi = \frac{J_m}{Y_m} (nr) \frac{\cos}{\sin} m\theta \frac{\cosh}{\sinh} nz, m = 0, 1, 2, \ldots.$$

In our problem we must discard the Y solution as $Y_m(r)$ is infinite when $r = 0$. So, choosing our zero of θ suitably, we can write

$$\phi = J_m(nr) \cos m\theta \, (A \cosh nz + B \sinh nz).$$

At the bottom of the tank condition (ii) gives, as in §55, $A \sinh nh = B \cosh nh$, so that

$$\phi = C J_m(nr) \cos m\theta \cosh n(z + h).$$

The constants m and n are not independent, since we have to satisfy the boundary condition at $r = a$. This gives $J'_m(na) = 0$, so that for any selected m, n is restricted to have one of a certain set of values, determined from the roots of the above equation. The function C above will involve the time, and in fact if we are interested in waves whose frequency is f, we shall try $C \propto \sin 2\pi f t$. Putting $C = D \sin 2\pi f t$, where D is now a constant independent of r, θ, z or t, we have

$$\phi = DJ_m(nr) \cos m\theta \cosh n(z+h) \sin 2\pi f t. \tag{39}$$

The boundary condition §54 (iii) now enables us to find ζ; it is

$$\zeta = \frac{2\pi Df}{g} J_m(nr) \cos m\theta \cosh nh \cos 2\pi f t. \tag{40}$$

The remaining boundary condition §54 (iv) gives us the period equation; it is

$$-4\pi^2 f^2 \, DJ_m(nr) \cos m\theta \cosh nh \sin 2\pi f t$$

$$+gnDJ_m(nr) \cos m\theta \sinh nh \sin 2\pi f t = 0,$$

or

$$4\pi^2 f^2 = gn \tanh nh. \tag{41}$$

For waves with a selected value of m (which must be integral) n is found and hence, from (41), f is found. We conclude that only certain frequencies are allowed. Apart from an arbitrary multiplicative constant, the nature of the waves is now completely determined.

§57   Paths of the particles

In §55 we discussed the progressive wave motion in an infinite straight channel. It is possible to determine from (34) the actual paths of the particles of fluid in this motion. For if X, Z denote the displacements of a particle whose mean position is (x, z) we have

$$\dot{X} = -\frac{\partial \phi}{\partial x} = \frac{ga}{c} \frac{\cosh m(z+h)}{\cosh mh} \sin m(x-ct),$$

$$\dot{Z} = -\frac{\partial \phi}{\partial z} = -\frac{ga}{c} \frac{\sinh m(z+h)}{\cosh mh} \cos m(x-ct),$$

in which we have neglected terms of the second order and a dot signifies differentiation with respect to time. Thus

$$X = \frac{ga \cosh m(z+h)}{mc^2 \cosh mh} \cos m(x-ct),$$

$$Z = \frac{ga \sinh m(z+h)}{mc^2 \cosh mh} \sin m(x-ct).$$

Eliminating t, we find for the required path

$$\frac{X^2}{\cosh^2 m(z+h)} + \frac{Z^2}{\sinh^2 m(z+h)} = \frac{g^2 a^2}{m^2 c^4 \cosh^2 mh}. \qquad (42)$$

These paths are ellipses in a vertical plane with a constant distance $(2ga/mc^2)$ sech mh between their foci. A similar discussion could be given for the other types of wave motion which we have solved in other paragraphs.

§58 The kinetic and potential energies

The kinetic (K.E.) and potential energies (P.E.) of these waves are easily determined. Thus, if we measure the P.E. relative to the undisturbed state, then, since $\zeta(x, y)$ is the elevation, the mass of liquid standing above a base δA in the (x, y) plane is $\rho\zeta\delta A$. Its centre of mass is at a height $\frac{1}{2}\zeta$, and thus the total P.E. is

$$\int \tfrac{1}{2} g\rho\zeta^2 \, dA, \qquad (43)$$

the integral being taken over the undisturbed area of surface. Likewise the K.E. of a small element is $\frac{1}{2}\rho\mathbf{u}^2 \, d\tau$, $d\tau$ being the element of volume of the liquid, so that the total K.E. is

$$T = \int \tfrac{1}{2} \rho\mathbf{u}^2 \, d\tau, \qquad (44)$$

the integral being taken over the whole liquid, which may, within our approximation, be taken to be the undisturbed volume.

With the progressive waves of §55, ζ and ϕ are given by (33) and (34), and a simple integration shows that the K.E. and P.E. in one wavelength $(2\pi/m)$ are equal, and per unit width of stream, have the value

$$\tfrac{1}{4} g\rho a^2 \lambda. \qquad (45)$$

In evaluating (44) it is often convenient to use Green's Theorem in the form

$$\iint \left\{ \left(\frac{\partial\phi}{\partial x}\right)^2 + \left(\frac{\partial\phi}{\partial y}\right)^2 + \left(\frac{\partial\phi}{\partial x}\right)^2 \right\} d\tau = \int \phi \frac{\partial\phi}{\partial\nu} \, dS.$$

The latter integral is taken over the surface S which bounds the original volume τ, and $\partial/\partial \nu$ represents differentiation along the outward normal to this volume. Since $\partial \phi/\partial \nu = 0$ on a fixed boundary, some of the contributions to T will generally vanish. Also, on the free surface, if ζ is small, we may put $\partial \phi/\partial z$ instead of $\partial \phi/\partial \nu$.

§59   Rate of transmission of energy

We shall next calculate the rate at which energy is transmitted in one of these surface waves. We can illustrate the method by considering the problem discussed in §55, concerning progressive waves in a rectangular tank of depth h. Let AA' in Fig. 20 be an imaginary plane fixed in the liquid perpendicular to the direction of wave propagation. We shall calculate the rate at which the liquid on the left of AA' is doing work

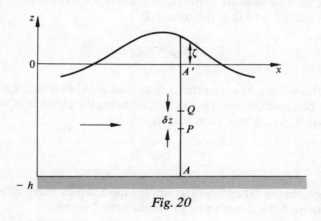

Fig. 20

upon the liquid on the right. This will represent the rate at which the energy is being transmitted. Suppose that the tank is of unit width and consider that part of AA' which lies between the two lines $z, z + \delta z$ (shown as PQ in the figure). At all points of this area the pressure is p, and the velocity is u. The rate at which work is being done is therefore $pu \, \delta z$. Thus the total rate is

$$\int_{-h}^{0} pu \, \delta z.$$

We use Bernoulli's equation (8) to give us p; since $\mathbf{u}^2$ may be neglected, and $V = gz$, therefore

$$p = p_0 + \rho \frac{\partial \phi}{\partial t} - g\rho z.$$

Now, according to (1) $u = -\partial\phi/\partial x$ and from (34),

$$\phi = \frac{ga}{mc} \frac{\cosh m(z+h)}{\cosh mh} \cos m(x-ct).$$

Putting these various values in the required integral we obtain

$$\sin m(x-ct) \int_{-h}^{0} \frac{ga}{c} \frac{\cosh m(z+h)}{\cosh mh} (p_0 - g\rho z)\, dz$$
$$+ \sin^2 m(x-ct) \int_{-h}^{0} \frac{\rho g^2 a^2}{c} \frac{\cosh^2 m(z+h)}{\cosh^2 mh}\, dz.$$

This expression fluctuates with the time, and we are concerned with its mean value over a cycle. The mean value of $\sin m(x-ct)$ is zero, and of $\sin^2 m(x-ct)$ is $\frac{1}{2}$. Thus the mean rate at which work is being done is

$$\frac{\rho g^2 a^2}{2c} \operatorname{sech}^2 mh \int_{-h}^{0} \cosh^2 m(z+h)\, dz,$$

which is easily seen to be

$$\tfrac{1}{4} g\rho a^2 c (1 + 2mh \operatorname{cosech} 2mh).$$

In terms of the wavelength $\lambda = 2\pi/m$, this is

$$\tfrac{1}{4} g\rho a^2 c \left\{ 1 + \frac{4\pi h}{\lambda} \operatorname{cosech} \frac{4\pi h}{\lambda} \right\}. \tag{46}$$

Now from (45) we see that the total energy with a stream of unit width is $\frac{1}{2} g\rho a^2$ per unit length. Thus the velocity of energy flow is

$$\frac{c}{2} \left\{ 1 + \frac{4\pi h}{\lambda} \operatorname{cosech} \frac{4\pi h}{\lambda} \right\}. \tag{47}$$

We shall see in a later chapter that this velocity is an important quantity known as the *group velocity*.

§60   Inclusion of surface tension
General formulae

In the preceding paragraphs we have assumed that surface tension could be neglected. However, with short waves this is not satisfactory and we must now investigate the effect of allowing for it. When we say that the surface tension is T, we mean that if a line of unit length is drawn in the surface of the liquid, then the liquid on one side of this line exerts a pull on the liquid on the other side, of magnitude T. Thus the effect of surface tension is similar to that of a membrane everywhere stretched to a tension T (as in Chapter 3, §32) placed on the surface of the liquid. We showed in Chapter 3 that when the membrane was bent there was a downward force per unit area approximately equal to

$$-T\left\{\frac{\partial^2 \zeta}{\partial x^2}+\frac{\partial^2 \zeta}{\partial y^2}\right\}.$$

Thus in Fig. 21 the pressure p_1 just inside the liquid does not equal the atmospheric pressure p_0, but rather

$$p_1 = p_0 - T\left\{\frac{\partial^2 \zeta}{\partial x^2}+\frac{\partial^2 \zeta}{\partial y^2}\right\}. \tag{48}$$

The reader who is familiar with hydrostatics will recognise that the excess pressure inside a stretched film (as in a soap bubble) is $T(1/R_1 + 1/R_2)$, where R_1 and R_2 are the radii of curvature in any pair of perpendicular planes through the normal to the surface. We may put $R_1 = -\partial^2 \zeta/\partial x^2$ and $R_2 = -\partial^2 \zeta/\partial y^2$ to the first order of small quantities, and then (48) follows immediately.

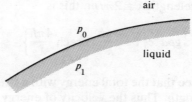

Fig. 21

Thus, instead of being $p = p_0$ at the free surface of the liquid, the correct condition is that $p + T\left\{\dfrac{\partial^2 \zeta}{\partial x^2}+\dfrac{\partial^2 \zeta}{\partial y^2}\right\}$ is constant and equal to p_0.

We may combine this with Bernoulli's equation (9), in which we neglect $\mathbf{u}^2$ and put $V = gz$. Then the new boundary condition which replaces §54 (iii) is now

$$\frac{\partial \phi}{\partial t} - g\zeta + \frac{\mathsf{T}}{\rho}\left\{ \frac{\partial^2 \zeta}{\partial x^2} + \frac{\partial^2 \zeta}{\partial y^2} \right\} = 0, \tag{49}$$

We still have the boundary condition §54 (iv) holding, since this is not affected by any sudden change in pressure at the surface. By combining (29) and (49) we find the new condition that replaces §54 (v). It is

$$\frac{\partial^2 \phi}{\partial t^2} + g\frac{\partial \phi}{\partial z} - \frac{\mathsf{T}}{\rho}\left\{ \frac{\partial^2}{\partial x^2} + \frac{\partial^2}{\partial y^2} \right\}\frac{\partial \phi}{\partial z} = 0. \tag{50}$$

We may collect these formulae together; thus, with surface tension

(i) $\nabla^2 \phi = 0$ in the body of the liquid $\hfill (4)$

(ii) $\partial \phi / \partial \nu = 0$ on all fixed boundaries $\hfill (2)$

(iii) $\dfrac{\partial \phi}{\partial t} - g\zeta + \dfrac{\mathsf{T}}{\rho}\left\{ \dfrac{\partial^2 \zeta}{\partial x^2} + \dfrac{\partial^2 \zeta}{\partial y^2} \right\} = 0$ on the free surface $\hfill (49)$

(iv) $\partial \zeta / \partial t = -\partial \phi / \partial z$ on the free surface $\hfill (29)$

(v) $\dfrac{\partial^2 \phi}{\partial t^2} + g\dfrac{\partial \phi}{\partial z} - \dfrac{\mathsf{T}}{\rho}\left\{ \dfrac{\partial^2}{\partial x^2} + \dfrac{\partial^2}{\partial y^2} \right\}\dfrac{\partial \phi}{\partial z} = 0$ on the free surface $\hfill (50)$

Only two of the last three equations are independent.

§61 Capillary waves in one dimension

Waves of the kind in which surface tension is important are known as **capillary waves**. We shall discuss one case which will illustrate the conditions (i)–(v). Let us consider progressive type waves on an unlimited sheet of water of depth h, assuming that the motion takes place exclusively in the direction of x. Then, by analogy with (31) we shall try

$$\phi = C \cosh m(z + h) \cos m(x - ct). \tag{51}$$

This satisfies (i) and (ii); (iv) gives the form of ζ, which is

$$\zeta = (C/c) \sinh mh \sin m(x - ct). \tag{52}$$

We have only one more condition to satisfy; if we choose (v) this gives

$$-m^2c^2C \cosh mh \cos m(x-ct) + mCg \sinh mh \cos m(x-ct)$$

$$+\frac{T}{\rho}m^3C \sinh mh \cos m(x-ct) = 0,$$

giving

$$c^2 = (g/m + Tm/\rho) \tanh mh. \tag{53}$$

This equation is really the modified version of (32) when allowance is made for the surface tension; if $T = 0$, it reduces to (32).

When h is large, $\tanh mh = 1$, and if we write $m = 2\pi/\lambda$, we have

$$c^2 = \frac{g\lambda}{2\pi} + \frac{2\pi T}{\lambda\rho}. \tag{54}$$

The curve of c against λ is shown in Fig. 22, from which it can be seen that there is a minimum velocity which occurs when $\lambda^2 = 4\pi^2 T/g\rho$. Waves shorter than this, in which surface tension is dominant, are called **ripples**, and it is seen that for any velocity greater than the minimum there are two possible types of progressive wave, one of which is a ripple. The minimum velocity is $(4gT/\rho)^{1/4}$, and in water this critical velocity is about $0\cdot23$ m sec^{-1}, and the critical wavelength is about $0\cdot017$ m. Curves of c against λ for other values of the depth h are very similar to those in Fig. 22.

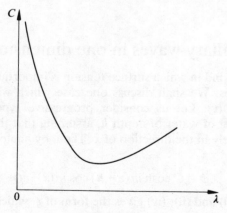

Fig. 22

§62 Examples

1. Find the potential and kinetic energies for tidal waves in a tank of length l, using the notation of §49.

2. Find the velocity of any particle of liquid in the problem of tidal waves in a circular tank of radius a (§51). Show that when $m = 0$ in (25), particles originally on a vertical cylinder of radius r coaxial with the tank, remain on a coaxial cylinder whose radius fluctuates; find an expression for the amplitude of oscillation of this radius in terms of r.

3. Tidal waves are occurring in a square tank of depth h and side a. Find the normal modes, and calculate the kinetic and potential energies for each of them. Show that when more than one such mode is present, the total energy is just the sum of the separate energies of each normal mode.

4. What are the paths of the particles of the fluid in the preceding question?

5. A channel of unit width is of depth h, where $h = kx$, k being a constant. Show that tidal waves are possible with frequency $p/2\pi$, for which

$$\zeta = AJ_0(ax^{1/2}) \cos pt,$$

where $a^2 = 4p^2/kg$, and J_0 is Bessel's function of order zero. It is known that the distance between successive zeros of $J_0(x)$ tends to π when x is large. Hence show that the wavelength of these stationary waves increases with increasing values of x. (This is the problem of a shelving beach.)

6. At the end of a shallow tank, we have $x = 0$, and the depth of water h is $h = h_0 x^{2m}$. Also the breadth of the tank b is given by $b = b_0 x^n$. Show that tidal waves of frequency $p/2\pi$ are possible, for which

$$\zeta = Ax^u J_q(rx^s) \cos pt,$$

where

$$s = 1 - m, \qquad a^2 = p^2/gh_0, \qquad r = a/s,$$
$$2u = 1 - 2m - n \quad \text{and} \quad q = |u/s|.$$

Use the fact that $J_m(x)$ satisfies the equation

$$\frac{d^2 J}{dx^2} + \frac{1}{x}\frac{dJ}{dx} + \left(1 - \frac{m^2}{x^2}\right)J = 0.$$

7. Prove directly from the conditions (i)–(v) in §54 without using the results of §55 that the velocity of surface waves in a rectangular channel of infinite depth is $\sqrt{(g\lambda/2\pi)}$.

8. Find the paths of particles of fluid in the case of surface waves on an infinitely deep circular tank of radius a.

9. A tank of depth h is in the form of a sector of a circle of radius a and angle 72°. What are the allowed normal modes for surface waves?

10. If X, Y, Z denotes the displacement of a particle of fluid from its mean position x, y, z in a rectangular tank of sides a and b when surface waves given by equation (37) are occurring, prove that the path of the particle is the straight line

$$\frac{a}{p\pi}\cot\frac{p\pi x}{a}X = \frac{b}{q\pi}\cot\frac{q\pi y}{b}Y = \frac{1}{r}\coth r(z+h)Z.$$

11. Show that in surface waves on a cylindrical tank of radius a and depth h, the energies given by the normal modes (39) are

$$V = \frac{2\pi^3 D^2 f^2 \rho}{g}\cosh^2 nh \cos^2 2\pi ft \int_0^a J_m^2(nr)\, r\, dr,$$

and

$$T = \tfrac{1}{2}n\pi\rho D^2 \sin^2 2\pi ft \cosh nh \sinh nh \int_0^a J_m^2(nr)\, r\, dr.$$

Use the fact that the total energy must be independent of the time to deduce from this that the period equation is

$$4\pi^2 f^2 = gn \tanh nh.$$

12. Show that when we use cylindrical polar coordinates to describe the capillary waves of §60, the pressure condition at the free surface §60 (iii) is

$$\frac{\partial \phi}{\partial t} - g\zeta + \frac{T}{\rho}\left\{\frac{\partial^2 \zeta}{\partial r^2} + \frac{1}{r}\frac{\partial \zeta}{\partial r} + \frac{1}{r^2}\frac{\partial^2 \zeta}{\partial \theta^2}\right\} = 0.$$

Use this result to show that waves of this nature on a circular basin of infinite depth are described by

$$\phi = CJ_m(nr)\cos m\theta \cos 2\pi ft,$$

$$\zeta = \frac{-nC}{2\pi f}J_m(nr)\cos m\theta \sin 2\pi ft,$$

where

$$J'_m(na) = 0 \quad \text{and} \quad 4\pi^2 f^2 = gn + \mathsf{T}n^3/\rho.$$

13. Show that capillary waves on a rectangular basin of sides a, b and depth h are given by

$$\phi = A \frac{\cosh r(z+h)}{\sinh rh} \cos \frac{m\pi x}{a} \cos \frac{n\pi y}{b} \cos 2\pi ft,$$

$$\zeta = -\frac{rA}{2\pi f} \cos \frac{m\pi x}{a} \cos \frac{n\pi y}{b} \sin 2\pi ft,$$

where $m = 0, 1, 2, \ldots$; $n = 0, 1, 2 \ldots$; $r^2 = \pi^2(m^2/a^2 + n^2/b^2)$, and the period equation is

$$4\pi^2 f^2 = (gr + \mathsf{T}r^3/\rho) \tanh rh.$$

Verify, that when $n = 0$, this is equivalent to the result of §61, equation (53).

Sound waves

§63 Relation between pressure and density

Throughout Chapter 5 we assumed that the liquid was incompressible. An important class of problems is that of waves in a compressible fluid, such as a gas. In this chapter we shall discuss such waves, of which sound waves are particular examples. The passage of a sound wave through a gas is accompanied by oscillatory motion of particles of the gas in the direction of wave propagation. These waves are therefore longitudinal. Since the density ρ is not constant, but varies with the pressure p, we require to know the relation between p and ρ. If the compressions and rarefactions that compose the wave succeed each other so slowly that the temperature remains constant (an *isothermal* change) this relation is $p = k\rho$. But normally this is not the case and no flow of heat, which would be needed to preserve the temperature constant, is possible; in such cases (*adiabatic* changes)

$$p = k\rho^{\gamma}, \tag{1}$$

where k and γ are constants depending on the particular gas used. We shall use (1) when it is required, rather than the isothermal relation.

§64 The governing differential equation

There are several problems in the propagation of sound waves that can be solved without using the apparatus of velocity potential ϕ in the form in which we used it in Chapter 5, §§54–61; we shall therefore discuss some of these before giving the general development of the subject.

Our first problem is that of waves along a uniform straight tube, or pipe, and we shall be able to solve this problem in a manner closely akin to that of Chapter 4, §38, where we discussed the longitudinal vibrations of a rod. We can suppose that the motion of the gas particles is entirely in the direction of the tube, and that the velocity and displacement are the same for all points of the same cross-section.

Suppose for convenience that the tube is of unit cross-sectional area, and let us consider the motion of that part of the gas originally confined between parallel planes at P and Q a distance δx apart as in Fig. 23.

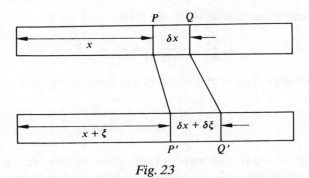

Fig. 23

The plane P is distant x from some fixed origin in the tube. During the vibration let PQ move to $P'Q'$, in which P is displaced a distance ξ from its mean position, and Q a distance $\xi + \delta\xi$. The length $P'Q'$ is therefore $\delta x + \delta\xi$. We shall find the equation of motion of the gas at $P'Q'$. For this purpose we shall require to know its mass and the pressure at its two ends. Its mass is the same as the mass of the undisturbed element PQ, namely, $\rho_0 x$, where ρ_0 is the normal average density. To get the pressure at P' we imagine the element δx to shrink to zero; this gives the local density ρ, from which, by (1), we calculate the pressure. We have

$$\rho = \mathop{\mathrm{Lim}}_{\delta x \to 0} \rho_0\, \delta x / (\delta x + \delta\xi) = \rho_0\left(1 + \frac{\partial\xi}{\partial x}\right)^{-1} = \rho_0\left(1 - \frac{\partial\xi}{\partial x}\right), \qquad (2)$$

if we may neglect powers of $\partial\xi/\partial x$ higher than the first. The quantity $(\rho - \rho_0)/\rho_0$ will often occur in this chapter; it is called the **condensation** s. Thus

$$s = -\partial\xi/\partial x, \qquad \rho = \rho_0(1 + s). \qquad (3)$$

The net force acting on the element $P'Q'$ is $p_{P'} - p_{Q'}$, and hence the equation of motion is

$$\rho_0\, \delta x\, \frac{\partial^2\xi}{\partial t^2} = p_{P'} - p_{Q'} = -\frac{\partial p}{\partial x}\,\delta x,$$

or,

$$\rho_0 \frac{\partial^2 \xi}{\partial t^2} = -\frac{\partial p}{\partial x}. \tag{4}$$

We may use (2) to rewrite (4) in the form

$$\rho_0 \frac{\partial^2 \xi}{\partial t^2} = -\frac{dp}{d\rho} \frac{\partial \rho}{\partial x} = \rho_0 \frac{dp}{d\rho} \frac{\partial^2 \xi}{\partial x^2}.$$

It thus appears then that ξ satisfies the familiar equation of wave motion

$$\frac{\partial^2 \xi}{\partial x^2} = \frac{1}{c^2} \frac{\partial^2 \xi}{\partial t^2} \quad \text{with } c^2 = dp/d\rho. \tag{5}$$

This equation shows that waves of any shape will be transmitted in either direction with velocity $\sqrt{(dp/d\rho)}$. In the case of ordinary air at $0°$ C, using (1) as the relation between p and ρ, we find that the velocity is $c = 332$ metres per sec., which agrees with experiment. Newton, who made this calculation originally, took the isothermal relation between p and ρ and, naturally, obtained an incorrect value for the velocity of sound.

A more accurate calculation of the equation of motion can be made, in which powers of $\partial \xi / \partial x$ are not neglected, as follows. From (2) we have the accurate result

$$p = k\rho^\gamma = k\rho_0^\gamma \left(1 + \frac{\partial \xi}{\partial x}\right)^\gamma.$$

So, now using (4) in which no approximations have been made,

$$\rho_0 \frac{\partial^2 \xi}{\partial t^2} = \frac{\gamma k \rho_0^\gamma}{\{1 + \partial \xi / \partial x\}^{\gamma+1}} \frac{\partial^2 \xi}{\partial x^2}$$

or,

$$\frac{\partial^2 \xi}{\partial t^2} = \frac{\gamma p_0}{\rho_0} \frac{1}{\{1 + \partial \xi / \partial x\}^{\gamma+1}} \frac{\partial^2 \xi}{\partial x^2}. \tag{6}$$

Equation (5) is found from (6) by neglecting $\partial \xi / \partial x$ compared with unity. A complete solution of (6) is, however, beyond the scope of this book. It is easy to see that, since (6) is not in the standard form of a wave equation, the velocity of transmission depends upon the frequency, and hence as the equation is dispersive a wave is not, in general, transmitted without change of shape.

§65 Solutions for a pipe of finite length

We must now discuss the boundary conditions. With an infinite tube, of course, there are no such conditions, but with a tube rigidly closed at $x = x_0$, we must have $\xi = 0$ at $x = x_0$, since at a fixed boundary the gas particles cannot move.

Another common type of boundary condition occurs when a tube has one or more ends open to the atmosphere. If we suppose that the waves inside the tube do not extend their influence to the air beyond the end of the tube, then at all open ends the pressure must have the normal atmospheric value, and thus, from (1) and (2), $\partial \xi / \partial x = 0$. Since the waves do extend a little outside the tube, this last equation is not strictly accurate. The usual modification is to increase the effective length of the tube by a small end-correction depending on the area of the cross-section of the tube. We shall not, however, include such corrections in what follows.

To summarise:

(i) $\dfrac{\partial^2 \xi}{\partial x^2} = \dfrac{1}{c^2} \dfrac{\partial^2 \xi}{\partial t^2}$ in the tube, and $c^2 = \mathrm{d}p/\mathrm{d}\rho$, $\qquad$ (5)

(ii) $\xi = 0$ at a closed end, $\qquad$ (7)

(iii) $-\dfrac{\partial \xi}{\partial x} = s = 0$ at an open end. $\qquad$ (8)

§66 Normal modes

We shall apply these equations to find the normal modes of vibration of gas in a tube of length l. These waves will naturally be of stationary type.

(a) *Closed at both ends* $x = 0, l$. This problem is the same mathematically as the transverse vibrations of a string of length l fixed at its ends (cf. Chapter 2, §23). Conditions (i) and (ii) of §65 give for the normal modes

$$\xi_r = A_r \sin \frac{r\pi x}{l} \cos \left\{ \frac{r\pi ct}{l} + \varepsilon_r \right\}, \qquad r = 1, 2, \ldots. \qquad (9)$$

(b) *Closed at* $x = 0$, *open at* $x = l$ (a "stopped tube"). Here conditions (ii) and (iii) give $\xi = 0$ at $x = 0$, and $\partial \xi / \partial x = 0$ at $x = l$. The normal modes are

$$\xi_r = A_r \sin\left(r+\frac{1}{2}\right)\frac{\pi x}{l} \cos\left\{\left(r+\frac{1}{2}\right)\frac{\pi c t}{l}+\varepsilon_r\right\}, \qquad r = 0, 1, 2, \ldots .$$

$$(10)$$

(c) *Open at both ends* $x = 0, l$. We have to satisfy the boundary condition (iii) $\partial\xi/\partial x = 0$ at $x = 0, l$. So the normal modes are

$$\xi_r = A_r \cos\frac{r\pi x}{l} \cos\left\{\frac{r\pi c t}{l}+\varepsilon_r\right\}, \qquad r = 1, 2, \ldots . \quad (11)$$

In each case the full solution would be the superposition of any number in terms of the appropriate type with different r. The fundamental periods in the three cases are $2l/c$, $4l/c$ and $2l/c$, respectively. The harmonics bear a simple numerical relationship to the fundamental, which explains the pleasant sound of an organ pipe.

§67   Normal modes in a tube with moveable boundary

We shall now solve a more complicated problem. We are to find the normal modes of a tube of unit sectional area, closed at one end by a rigid boundary and at the other by a mass M free to move along the tube. Let the fixed boundary be taken as $x = 0$, and the normal equilibrium position of the moveable mass be at $x = l$ as in Fig. 24.

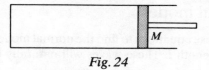

Fig. 24

Then we have to solve the standard wave equation with the boundary conditions that when $x = 0$, (ii) gives $\xi = 0$, and that when $x = l$ the excess pressure inside, $p - p_0$, must be responsible for the acceleration of the mass M. This implies that

$$p - p_0 = M\frac{\partial^2\xi}{\partial t^2} \quad \text{when } x = l.$$

The first condition is satisfied by the function

$$\xi = A \sin nx \cos(nct+\varepsilon). \qquad (12)$$

To satisfy the second condition, we observe that from (3),

$$p - p_0 = (dp/d\rho)(\rho - \rho_0) = -c^2 \rho_0 \, \partial\xi/\partial x.$$

So this condition becomes

$$M\frac{\partial^2\xi}{\partial t^2} = -c^2\rho_0\frac{\partial\xi}{\partial x} \quad \text{at } x = l.$$

Using (12) this gives, after a little reduction,

$$nl \tan nl = l\rho_0/M.$$

The allowed values of n are the roots of this equation. There is an infinite number of them, and when $M = 0$, so that the tube is effectively open to the air at one end, we obtain equation (10); when $M = \infty$, so that the tube is closed at each end, we obtain equation (9).

§68　The velocity potential. General formulae

So far we have developed our solutions in terms of ξ, the displacement of any particle of the gas from its mean position. It is possible, however, to use the method of the velocity potential ϕ. Many of the conditions which ϕ must satisfy are the same as in Chapter 5, but a few of them are changed to allow for the variation in density. It is convenient to gather these various formulae together first.

  (i) If the motion is irrotational, as we shall assume, $\mathbf{u} = -\nabla\phi$, (cf. Chapter 5, equation (1));   (13)
 (ii) at any fixed boundary, $\partial\phi/\partial\nu = 0$ (cf. Chapter 5, equation (2));(14)
(iii) the equation of continuity (cf. Chapter 5, equation (3)) is slightly altered, and it is now

$$\frac{\partial\rho}{\partial t} + \operatorname{div}(\rho\mathbf{u}) = 0,$$

or

$$\frac{\partial\rho}{\partial t} + \frac{\partial}{\partial x}(\rho u) + \frac{\partial}{\partial y}(\rho v) + \frac{\partial}{\partial z}(\rho w) = 0. \tag{15}$$

(iv) The equations of motion are unchanged; if $\mathbf{F}$ is the external force on unit mass, in vector form, they are

$$\frac{D\mathbf{u}}{Dt} = \mathbf{F} - \frac{1}{\rho}\nabla p \text{ (cf. Chapter 5, equation (6))}. \tag{16}$$

(v) In cases where the external forces have a potential V, we obtain Bernoulli's equation (cf. Chapter 5, equation (8))

$$\int \frac{\mathrm{d}p}{\rho} + \frac{1}{2}\mathbf{u}^2 + V - \frac{\partial \phi}{\partial t} = \text{const.} \tag{17}$$

in which we have absorbed an arbitrary function of the time into the term $\partial \phi / \partial t$ (cf. Chapter 5, equation (8)).

§69   The differential equation of wave motion

In sound waves we may neglect all external forces except such as occur at boundaries, and thus we may put $V = 0$ in (17). Also we may suppose that the velocities are small and neglect $\mathbf{u}^2$ in this equation. With these approximations Bernoulli's equation becomes

$$\int \frac{\mathrm{d}p}{\rho} - \frac{\partial \phi}{\partial t} = \text{const.}$$

We can simplify the first term; for

$$\int \frac{\mathrm{d}p}{\rho} = \int \left(\frac{\mathrm{d}p}{\mathrm{d}\rho}\right)\frac{\mathrm{d}\rho}{\rho},$$

and if the variations in density are small, $\mathrm{d}p/\mathrm{d}\rho$ may be taken as constant, and equal to c^2 as in (5). Thus

$$\int \frac{\mathrm{d}p}{\rho} = c^2 \int \frac{\mathrm{d}\rho}{\rho} = c^2 \log_e \rho = c^2 \{\log_e (1+s) + \log_e \rho_0\}.$$

So

$$\int \frac{\mathrm{d}p}{\rho} = c^2 s + \text{const.},$$

if s is small. If we absorb this constant in ϕ, then Bernoulli's equation takes its final form

$$c^2 s - \partial \phi / \partial t = 0. \tag{18}$$

Laplace's equation for ϕ does not hold because of the changed equation of continuity. But if u, v, w and s are small, (15) can be written in a simpler form by the aid of (13); namely,

$$\rho_0 \partial s / \partial t - \rho \nabla^2 \phi = 0.$$

This is effectively the same as

$$\frac{\partial s}{\partial t} = \nabla^2 \phi. \tag{19}$$

Now let us eliminate s between (18) and (19), and we shall find the standard of wave equation

$$\nabla^2 \phi = \frac{1}{c^2} \frac{\partial^2 \phi}{\partial t^2}. \tag{20}$$

This shows that c is indeed the velocity of wave propagation, but before we can use this technique for solving problems, we must first obtain the boundary conditions for ϕ. At a fixed boundary, by (ii) $\partial \phi / \partial \nu = 0$. At an open end of a tube, the pressure must be atmospheric, and hence $s = 0$. Thus, from (18),

$$\partial \phi / \partial t = 0. \tag{21}$$

This completes the development of the method of the velocity potential, and we can choose in any particular problem whether we solve by means of the displacement ξ or the potential ϕ. It is possible to pass from one to the other, since from (3) and (18)

$$\frac{\partial \xi}{\partial x} = -s = -\frac{1}{c^2} \frac{\partial \phi}{\partial t}. \tag{22}$$

§70   An example

We shall illustrate these equations by solving the problem of stationary waves in a tube of length l, closed at one end $(x = 0)$ and open at the other $(x = l)$. This is the problem already dealt with in §66(b), and with the same notation, we require a solution of

$$\frac{\partial^2 \phi}{\partial x^2} = \frac{1}{c^2} \frac{\partial^2 \phi}{\partial t^2},$$

subject to the conditions

$$\partial \phi / \partial x = 0 \text{ at } x = 0,$$
$$\partial \phi / \partial t = 0 \text{ at } x = l.$$

It is easily seen that

$$\phi = a \cos mx \cos (cmt + \varepsilon)$$

satisfies all these conditions provided that $\cos ml = 0$, and so $ml = \pi/2$, $3\pi/2, \ldots, (r+\frac{1}{2})\pi, \ldots$. So the normal modes are

$$\phi_r = a_r \cos\left(r+\frac{1}{2}\right)\frac{\pi x}{l} \cos\left\{\left(r+\frac{1}{2}\right)\frac{\pi ct}{l}+\varepsilon_r\right\},$$

and from this expression all the other properties of these waves may easily be obtained. The reader is advised to treat the problems of §66(a) and (c) in a similar manner.

§71   Spherical symmetry

Our next application of the equations of §69 will be to problems where there is spherical symmetry about the origin. The fundamental wave equation then becomes (see Chapter 1, equation (23))

$$\frac{\partial^2 \phi}{\partial r^2}+\frac{2}{s}\frac{\partial \phi}{\partial r}=\frac{1}{c^2}\frac{\partial^2 \phi}{\partial t^2},$$

with solutions of progressive type

$$\phi = \frac{1}{r}f(r-ct)+\frac{1}{r}g(r+ct).$$

There are solutions of stationary type (see Chapter 1, equation (37))

$$\phi = r^{-1}\frac{\cos}{\sin}\,mr\,\frac{\cos}{\sin}\,cmt.$$

If the gas is contained inside a fixed sphere of radius a, then we must have ϕ finite when $r = 0$, and $\partial\phi/\partial r = 0$ when $r = a$. This means that

$$\phi = \frac{A}{r}\sin mr \cos(cmt+\varepsilon),$$

with the condition

$$\tan ma = ma. \tag{23}$$

This period equation has an infinite number of roots which approximate to $ma = (n+1/2)\pi$ when n is large. So for its higher frequencies the system behaves very like a uniform pipe of length a open at one end and closed at the other.

This analysis would evidently equally well apply to describe waves in a conical pipe.

§72   The kinetic and potential energies

We shall now calculate the energy in a sound wave. The kinetic energy is clearly $\int \frac{1}{2}\rho \mathbf{u}^2 \, dV$, where dV is an element of volume. We may put $\rho = \rho_0$ without loss of accuracy. In terms of the velocity potential this may be written

$$\int \frac{1}{2}\rho_0 (\nabla\phi)^2 \, dV = -\frac{1}{2}\rho_0 \int \phi \nabla^2\phi \, dV + \frac{1}{2}\rho_0 \int \phi \frac{\partial\phi}{\partial\nu} \, dS. \qquad (24)$$

The last expression follows from Green's theorem just as in Chapter 5, **§58**, and the surface integral is taken over the boundary of the gas. There is also potential energy because each small volume of gas is compressed or rarified, and work is stored up in the process. To calculate it, consider a small volume V_0, which during the passage of a wave is changed to V_1. If s_1 is the corresponding value of the condensation, then from (3), we have, to the first degree in s_1,

$$V_1 = V_0(1 - s_1). \qquad (25)$$

Further, suppose that during the process of compression, V and s are simultaneous intermediary values. Then we can write the work done in compressing the volume from V_0 to V_1 in the form

$$-\int_{V_0}^{V_1} p \, dV.$$

But, just as in (25), $V = V_0(1 - s)$, and hence

$$dV = -V_0 \, ds.$$

We may also write

$$p = p_0 + (dp/d\rho)(\rho - \rho_0)$$

$$= p_0 + c^2 \rho_0 s.$$

Thus the potential energy may be written

$$\int_0^{s_1} (p_0 + c^2 \rho_0 s) V_0 \, ds = p_0 V_0 s_1 + \frac{1}{2} c^2 \rho_0 V_0 s_1^2$$

$$= p_0(V_0 - V_1) + \frac{1}{2} c^2 \rho_0 V_0 s_1^2.$$

This is the contribution to the P.E. which arises from the volume V_0. The total P.E. may be found by integration. The first term will vanish in this process since it merely represents the total change in volume of the gas, which we may suppose to be zero. We conclude, therefore, that the potential energy is

$$\int \tfrac{1}{2} c^2 \rho_0 s^2 \, dV. \qquad (26)$$

It can easily be shown that with a progressive wave the K.E. and P.E. are equal; this does not hold for stationary waves, for which their sum remains constant.

§73   Progressive waves in a tube of varying section

We conclude this chapter with a discussion of the propagation of waves along a pipe whose cross-sectional area A varies slowly along its length. Our discussion is similar in many respects to the analysis in §64.

Consider the pipe shown in Fig. 25, and let us measure distances x along the central line. It will be approximately true to say that the velocity u is constant across any section perpendicular to the x axis. Suppose that the gas originally confined between the two planes P, Q

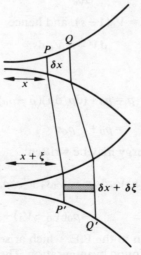

Fig. 25

at distances x, $x + \delta x$ is displaced during the passage of a wave, to $P'Q'$, the displacement of P being ξ and of Q being $\xi + \delta\xi$. Consider the motion of a small prism of gas such as that shaded in the figure; its equation of motion may be found as in **§64**, and it is

$$\rho_0 \frac{\partial^2 \xi}{\partial t^2} = -\frac{\partial p}{\partial x}. \tag{27}$$

We must therefore find the pressure in terms of ξ. This may be obtained from the equation of continuity, which expresses the fact that the mass of gas in $P'Q'$ is the same as that in PQ. Thus, if ρ is the density,

$$\rho_0 A(x)\, \delta x = \rho A(x + \xi) \cdot \{\delta x + \delta\xi\},$$

and so

$$\rho_0 A(x) = \rho \left\{ A(x) + \xi \frac{\partial A}{\partial x} \right\} \left\{ 1 + \frac{\partial \xi}{\partial x} \right\}.$$

Neglecting small quantities, this yields

$$\rho_0 = \rho \left\{ 1 + \frac{\partial \xi}{\partial x} + \frac{\xi}{A} \frac{\partial A}{\partial x} \right\}.$$

Therefore

$$\rho = \rho_0 \left\{ 1 - \frac{\partial \xi}{\partial x} - \frac{\xi}{A} \frac{\partial A}{\partial x} \right\} = \rho_0 \left\{ 1 - \frac{1}{A} \frac{\partial}{\partial x}(A\xi) \right\}. \tag{28}$$

Eliminating p between (27) and (28) we find

$$\rho_0 \frac{\partial^2 \xi}{\partial t^2} = -\frac{dp}{d\rho} \frac{\partial \rho}{\partial x} = c^2 \rho_0 \frac{\partial}{\partial x} \left\{ \frac{1}{A} \frac{\partial}{\partial x}(A\xi) \right\},$$

where, as usual, $c^2 = dp/d\rho$. So the equation of motion is

$$\frac{\partial^2 \xi}{\partial t^2} = c^2 \frac{\partial}{\partial x} \left\{ \frac{1}{A} \frac{\partial}{\partial x}(A\xi) \right\}. \tag{29}$$

In the case in which A is constant this reduces to the former equation (5). An important example when A is not constant is the co-called *exponential horn* used in the construction of high-fidelity tweeter speakers. In these the tube is approximately symmetrical about its central line and the area varies with the distance according to the law $A = A_0\, e^{2ax}$, where a and A_0 are constants.

With this form of A, (29) reduces to

$$\frac{\partial^2 \xi}{\partial t^2} = c^2 \left\{ \frac{\partial^2 \xi}{\partial x^2} + 2a \frac{\partial \xi}{\partial x} \right\}.$$

A solution is possible by the method of separation of variables (see §7). We find that

$$\xi = B_1 \exp(inct + m_1 x) + B_2 \exp(inct + m_2 x), \qquad (30)$$

where m_1 and m_2 are given by $-a \pm \sqrt{(a^2 - n^2)}$. Consequently, the general solution has the form

$$\xi = e^{-ax} \{ B_1 \exp[i(nct - (n^2 - a^2)^{1/2} x)]$$
$$+ B_2 \exp[i(nct + (n^2 - a^2)^{1/2} x)] \},$$

in which the first term represents a wave moving to the right while the second represents a wave moving to the left. These waves move along the horn at a speed c_h given by

$$c_h = c/(1 - a^2 c^2/\omega^2)^{1/2}, \qquad (31)$$

where we have set $\omega = nc$.

This expression indicates that propagation will only occur when the frequency $\omega > ac$. The quantity ac thus defines a **cutoff frequency** for the horn. We conclude from this that waves can be sent outwards along the horn with a speed which is approximately independent of the frequency, and with an attenuation factor e^{-ax} which is independent of the frequency. It is the approximate double independence of frequency which allows good reproduction of whatever waves are generated at the narrow end of the horn, and which is responsible for this choice of shape. Other forms of A will not, in general, give rise to the same behaviour.

§74   Examples

1. Use the method of §65 to investigate sound waves in a closed rectangular box of sides a_1, a_2 and a_3. Show that if the box is large, the number of such waves for which the frequency is less than n is approximately equal to one-eighth of the volume of the quadric $x^2/a_1^2 + y^2/a_2^2 + z^2/a_3^2 = 4n^2/c^2$. Hence show that this number is approximately $4\pi n^3 a_1 a_2 a_3/3c^3$.

2. Investigate the reflection and transmission of a train of harmonic waves in a uniform straight tube at a point where a smooth piston of mass M just fits into the tube and is free to move.

3. Show that the kinetic and potential energies of a plane progressive wave are equal.

4. Show that the kinetic and potential energies of stationary waves in a rectangular box have a constant sum.

5. Find an equation for the normal modes of a gas which is confined between two rigid concentric spheres of radii a and b.

6. Show that a closer approximation to the roots of equation (23) is $ma = (n + \frac{1}{2})\pi - 1/\{(n + \frac{1}{2})\pi\}$.

7. Find numerically the fundamental frequency of a conical pipe of radius one metre open at its wide end.

8. The cross-sectional area of a closed tube varies with the distance along its central line according to the law $A = A_0 x^n$. Show that if its two ends are $x = 0$, and $x = l$, then standing waves can exist in the tube for which the displacement is given by the formula.

$$\xi = x^{(1-n)/2} J_m(qx/c) \cos\{qct + \varepsilon\},$$

where

$$m = (n + 1)/2 \quad \text{and} \quad J_m(ql/c) = 0.$$

Use the fact that $J_m(x)$ satisfies the equation

$$\frac{d^2 J}{dx^2} + \frac{1}{x} \frac{dJ}{dx} + \left(1 - \frac{m^2}{x^2}\right) J = 0.$$

7

Electric waves

§75   Maxwell's equations

Before we discuss the propagation of electric waves, we shall summarise the most important equations that we shall require. These are known as **Maxwell's equations**. Let the vectors **E** (components E_x, E_y, E_z) and **H** (components H_x, H_y, H_z) denote the **electric** and **magnetic field strengths**. We shall suppose that all our media are isotropic with no ferromagnetism or permanent polarisation. Thus, if ε_0 and μ_0 are the **free space dielectric constant** and **permeability**, respectively, and we write $\varkappa_e$ for the **relative dielectric constant** (or **permittivity**) and $\varkappa_m$ for the **relative permeability**, the related vectors comprising the **magnetic induction B** and the **dielectric displacement D** are given by the equation $\mathbf{B} = \mu_0 \varkappa_m \mathbf{H}$, $\mathbf{D} = \varepsilon_0 \varkappa_e \mathbf{E}$. Further let **j** (components j_x, j_y, j_z) denote the **current density** vector, and ρ the **charge density**. Then, if we work in SI units with c denoting the velocity of light, Maxwell's equations may be summarised in vector form as follows:

$$\operatorname{div} \mathbf{D} = \rho, \tag{1}$$

$$\operatorname{div} \mathbf{B} = 0, \tag{2}$$

$$\operatorname{curl} \mathbf{H} = \mathbf{j} + \frac{\partial \mathbf{D}}{\partial t}, \tag{3}$$

$$\operatorname{curl} \mathbf{E} = -\frac{\partial \mathbf{B}}{\partial t} \tag{4}$$

where

$$\mathbf{D} = \varepsilon_0 \varkappa_e \mathbf{E}, \tag{5}$$

$$\mathbf{B} = \mu_0 \varkappa_m \mathbf{H}. \tag{6}$$

To these equations we must add the relation between **j** and **E**. If σ is the **conductivity**, which is the inverse of the specific resistance, this relation is

$$\mathbf{j} = \sigma \mathbf{E}. \qquad (7)$$

For conductors σ is large, and for insulators it is small.

The above equations have been written in vector form, but it is informative to display them also in terms of their components. If we wish to write these equations in their full Cartesian form, we have to remember that

$$\text{div } \mathbf{D} \equiv \nabla \cdot \mathbf{D} = \frac{\partial D_x}{\partial x} + \frac{\partial D_y}{\partial y} + \frac{\partial D_z}{\partial z}$$

and that

$$\text{curl } \mathbf{H} \equiv \nabla \times \mathbf{H} = \left(\frac{\partial H_z}{\partial y} - \frac{\partial H_y}{\partial z}, \frac{\partial H_x}{\partial z} - \frac{\partial H_z}{\partial x}, \frac{\partial H_y}{\partial x} - \frac{\partial H_x}{\partial y} \right).$$

The preceding equations then become

$$\frac{\partial D_x}{\partial x} + \frac{\partial D_y}{\partial y} + \frac{\partial D_z}{\partial z} = \rho \qquad (1) \qquad \frac{\partial B_x}{\partial x} + \frac{\partial B_y}{\partial y} + \frac{\partial B_z}{\partial z} = 0 \qquad (2')$$

$$\frac{\partial H_z}{\partial y} - \frac{\partial H_y}{\partial z} = j_x + \frac{\partial D_x}{\partial t} \qquad\qquad \frac{\partial E_z}{\partial y} - \frac{\partial E_y}{\partial z} = -\frac{\partial B_x}{\partial t}$$

$$\frac{\partial H_x}{\partial z} - \frac{\partial H_z}{\partial x} = j_y + \frac{\partial D_y}{\partial t} \qquad (3') \qquad \frac{\partial E_x}{\partial z} - \frac{\partial E_z}{\partial x} = -\frac{\partial B_y}{\partial t} \qquad (4')$$

$$\frac{\partial H_y}{\partial x} - \frac{\partial H_x}{\partial y} = j_z + \frac{\partial D_z}{\partial t} \qquad\qquad \frac{\partial E_y}{\partial x} - \frac{\partial E_x}{\partial y} = -\frac{\partial B_z}{\partial t}$$

$$D_x = \varepsilon_0 \varkappa_e E_x, \qquad D_y = \varepsilon_0 \varkappa_e E_y, \qquad D_z = \varepsilon_0 \varkappa_e E_z \qquad (5')$$

$$B_x = \mu_0 \varkappa_m H_x, \qquad B_y = \mu_0 \varkappa_m H_y, \qquad B_z = \mu_0 \varkappa_m H_z \qquad (6')$$

$$j_x = \sigma E_x, \qquad j_y = \sigma E_y, \qquad j_z = \sigma E_z. \qquad (7')$$

Equations (1)–(4) are sometimes called **Maxwell's Equations** and equations (5)–(7) **constitutive relations**. Simple physical bases can easily be given for (1)–(4). Thus, (1) represents Gauss' Theorem, and follows from the law of force between two charges; (2) represents the fact that isolated magnetic poles cannot be obtained; (3) is Ampere's Rule that the work done in carrying a unit pole round a closed circuit equals the total current enclosed in the circuit; part of this current is the conduction current $\mathbf{j}$ and part is Maxwell's displacement current $\dfrac{\partial \mathbf{D}}{\partial t}$; (4) is Lenz's law of induction.

These seven equations represent the basis of our subsequent work. They need to be supplemented by a statement of the boundary conditions that hold at a change of medium. If suffix n denotes the component normal to the boundary of the two media, and suffix s denotes the component in any direction in the boundary plane, then in passing from the one medium to the other

$$D_\text{n},\ B_\text{n},\ E_\text{s} \text{ and } H_\text{s} \text{ are continuous.} \tag{8}$$

In cases where there is a current sheet (i.e. a finite current flowing in an indefinitely thin surface layer) some of these conditions need modification, but we shall not discuss any such cases in this chapter.

There are two other important results that we shall use. First, we may suppose that the electromagnetic field stores energy, and the density of this energy per unit volume of the medium is

$$\tfrac{1}{2}(\varepsilon_0 \varkappa_\text{e} \mathbf{E}^2 + \mu_0 \varkappa_\text{m} \mathbf{H}^2). \tag{9}$$

Second, there is a vector, known as the **Poynting vector**, which is concerned with the rate at which energy is flowing. This vector, whose magnitude and direction are given by

$$(\mathbf{E} \times \mathbf{H}), \tag{10}$$

represents the amount of energy which flows in unit time across unit area drawn perpendicular to it. $\mathbf{E}$ and $\mathbf{H}$ are generally rapidly varying quantities and in such cases it is the mean value of (10) that has physical significance.

§76   Non-conducting media and the wave equation

We shall first deal with non-conducting media, such as glass, so that we may put $\sigma = 0$ in (7); we suppose that the medium is homogeneous, so that $\varkappa_\text{e}$ and $\varkappa_\text{m}$ are constants. If, as usually happens, there is no residual charge, we may also put $\rho = 0$ in (1), and with these simplifications,

Maxwell's equations may be written

$$\text{div } \mathbf{E} = 0, \qquad \text{div } \mathbf{H} = 0,$$

$$\text{curl } \mathbf{E} = -\mu_0 \varkappa_m \frac{\partial \mathbf{H}}{\partial t}, \qquad \text{curl } \mathbf{H} = \varepsilon_0 \varkappa_e \frac{\partial \mathbf{E}}{\partial t}. \tag{11}$$

These equations lead immediately to the standard wave equation, as we see by employing the identity

$$\text{curl curl } \mathbf{H} = \text{grad div } \mathbf{H} - \nabla^2 \mathbf{H}.$$

From the fourth of the equations in (11) we find

$$\text{grad div } \mathbf{H} - \nabla^2 \mathbf{H} = \varepsilon_0 \varkappa_e \text{ curl } \frac{\partial \mathbf{E}}{\partial t} = \varepsilon_0 \varkappa_e \frac{\partial}{\partial t} \text{ curl } \mathbf{E}.$$

Substituting for div $\mathbf{H}$ and curl $\mathbf{E}$, we discover the standard wave equation

$$\nabla^2 \mathbf{H} = \frac{\varkappa_e \varkappa_m}{c^2} \frac{\partial^2 \mathbf{H}}{\partial t^2}, \tag{12}$$

where we have set $\varepsilon_0 \mu_0 = 1/c^2$. Eliminating $\mathbf{H}$ instead of $\mathbf{E}$ we find the same equation for $\mathbf{E}$,

$$\nabla^2 \mathbf{E} = \frac{\varkappa_e \varkappa_m}{c^2} \frac{\partial^2 \mathbf{E}}{\partial t^2}. \tag{13}$$

According to our discussion of this equation in Chapter 1 this shows that waves can be propagated in such a medium, and that their velocity is $c/\sqrt{(\varkappa_e \varkappa_m)}$. In free space, where $\varkappa_e = \varkappa_m = 1$, this velocity is just c. But it is known that the velocity of light in free space has exactly this same value, so that we are thus led to the conclusion that light waves are electromagnetic in nature. X-rays, γ-rays, ultra-violet waves, infra-red waves and radio waves are also electromagnetic, and differ only in the order of magnitude of their wavelengths. We shall be able to show later, in §**78**, that these waves are transverse. Experiment has shown that $c = 2 \cdot 998 \times 10^8$ m/sec.

In non-conducting dielectric media, like glass, $\varkappa_e$ is not equal to unity; also $\varkappa_m$ depends on the frequency of the waves, but for light waves in the visible region we may put $\varkappa_m = 1$. The velocity of light is therefore $c/\sqrt{\varkappa_e}$. Now in a medium whose refractive index is K, it is known experimentally that the velocity of light is c/K. Hence, if our original assumptions are valid, $\varkappa_e = K^2$. This is known as **Maxwell's**

relation. It holds good for many substances, but fails because it does not take sufficiently detailed account of the atomic structure of the dielectric. It applies better for long waves (low frequency) than for short waves (high frequency).

§77   Electric and magnetic potentials

A somewhat different discussion of (11) can be given in terms of the electric and magnetic potentials. Since div $\mathbf{B} = 0$, it follows that we can write

$$\mathbf{B} = \mu_0 \varkappa_m \mathbf{H} = \text{curl } \mathbf{A}, \tag{14}$$

where $\mathbf{A}$ is a vector yet to be determined. This equation does not define $\mathbf{A}$ completely, since if ψ is any scalar, curl $(\mathbf{A} + \text{grad } \psi) = \text{curl } \mathbf{A}$. Thus $\mathbf{A}$ is undefined to the extent of addition of the gradient of any scalar, and we may accordingly impose one further condition upon it.

If $\mathbf{B} = \text{curl } \mathbf{A}$, and curl $\mathbf{E} = -\partial \mathbf{B}/\partial t$, it follows, by elimination of $\mathbf{B}$, that

$$\text{curl} \left\{ \mathbf{E} + \frac{\partial \mathbf{A}}{\partial t} \right\} = \mathbf{0}.$$

Integrating then gives

$$\mathbf{E} + \frac{\partial \mathbf{A}}{\partial t} = -\text{grad } \phi,$$

where ϕ is an arbitrary scalar function. Consequently we have

$$\mathbf{E} = -\text{grad } \phi - \frac{\partial \mathbf{A}}{\partial t}. \tag{15}$$

In cases where there is no variation with the time, this becomes $\mathbf{E} = -\text{grad } \phi$, showing that ϕ is the analogue of the electrostatic potential.

Eliminating $\mathbf{H}$ from the relations

$$\mu_0 \varkappa_m \mathbf{H} = \text{curl } \mathbf{A},$$

$$\text{curl } \mathbf{H} = \varepsilon_0 \varkappa_e \frac{\partial \mathbf{E}}{\partial t},$$

and using (15) to eliminate $\mathbf{E}$, we find

$$\text{grad div } \mathbf{A} - \nabla^2 \mathbf{A} = -\varepsilon_0 \mu_0 \varkappa_e \varkappa_m \left\{ \text{grad } \frac{\partial \phi}{\partial t} + \frac{\partial^2 \mathbf{A}}{\partial t^2} \right\},$$

and so

$$\nabla^2 \mathbf{A} = \frac{\varkappa_e \varkappa_m}{c^2} \frac{\partial^2 \mathbf{A}}{\partial t^2} + \text{grad} \left\{ \text{div } \mathbf{A} + \frac{\varkappa_e \varkappa_m}{c^2} \frac{\partial \phi}{\partial t} \right\},$$

where we have again used the fact that in SI units $\varepsilon_0 \mu_0 = 1/c^2$. Let us now introduce the extra allowed condition upon $\mathbf{A}$, and write

$$\text{div } \mathbf{A} + \frac{\varkappa_e \varkappa_m}{c^2} \frac{\partial \phi}{\partial t} = 0. \tag{16}$$

Then $\mathbf{A}$ satisfies the standard wave equation

$$\nabla^2 \mathbf{A} = \frac{\varkappa_e \varkappa_m}{c^2} \frac{\partial^2 \mathbf{A}}{\partial t^2}. \tag{17}$$

Further, taking the divergence of (15), we obtain, by (16)

$$0 = \text{div } \mathbf{E} = -\nabla^2 \phi - \frac{\partial}{\partial t} \text{div } \mathbf{A} = -\nabla^2 \phi + \frac{\varkappa_e \varkappa_m}{c^2} \frac{\partial^2 \phi}{\partial t^2}.$$

Thus ϕ also satisfies the standard wave equation

$$\nabla^2 \phi = \frac{\varkappa_e \varkappa_m}{c^2} \frac{\partial^2 \phi}{\partial t^2}. \tag{18}$$

A similar analysis can be carried through when ρ and j are not put equal to zero, and we find

$$\mathbf{B} = \text{curl } \mathbf{A}, \tag{14'}$$

$$\mathbf{E} = -\text{grad } \phi - \frac{\partial \mathbf{A}}{\partial t}, \tag{15'}$$

$$0 = \text{div } \mathbf{A} + \frac{\varkappa_e \varkappa_m}{c} \frac{\partial \phi}{\partial t}, \tag{16'}$$

$$\nabla^2 \mathbf{A} = \frac{\varkappa_e \varkappa_m}{c^2} \frac{\partial^2 \mathbf{A}}{\partial t^2} - \mu_0 \varkappa_m \mathbf{j}, \tag{17'}$$

$$\nabla^2 \phi = \frac{\varkappa_e \varkappa_m}{c^2} \frac{\partial^2 \phi}{\partial t^2} - \frac{\rho}{\varepsilon_0 \varkappa_e}. \tag{18'}$$

ϕ and $\mathbf{A}$ are known as the **electric potential** and **magnetic** or **vector potential**, respectively. It is open to our choice whether we solve problems in terms of $\mathbf{A}$ and ϕ, or of $\mathbf{E}$ and $\mathbf{B}$. The relations (14')–(18') enable us to pass from the one system to the other. The boundary conditions for ϕ and $\mathbf{A}$ may easily be obtained from (8), but since we shall always adopt the $\mathbf{E}$, $\mathbf{B}$ type of solution, which is usually the simpler, there is no need to write them down here.

There is one other general deduction that can be made here. If we use (3), (5) and (7) we can write, for homogeneous media

$$\text{curl } \mathbf{H} = \sigma \mathbf{E} + \varepsilon_0 \varkappa_e \frac{\partial \mathbf{E}}{\partial t}.$$

Taking the divergence of each side, and noting, from (1), that div $\mathbf{E} = \rho/\varepsilon_0 \varkappa_e$, we find

$$\frac{\partial \rho}{\partial t} + \frac{\sigma \rho}{\varepsilon_0 \varkappa_e} = 0.$$

Thus, on integration,

$$\rho = \rho_0 \, e^{-t/\theta}, \quad \text{where } \theta = \sigma/\varepsilon_0 \varkappa_e. \tag{19}$$

θ is called the **time of relaxation**. It follows from (19) that any original distribution of charge decays exponentially at a rate quite independent of any other electromagnetic disturbances that may be taking place simultaneously, and it justifies us in putting $\rho = 0$ in most of our problems. With metals such as copper, θ is of the order of 10^{-13} secs., and is beyond measurement; but with dielectrics such as water θ is large enough to be determined experimentally. Equation (19) only applies to the charge at an internal point in a medium; charges at the boundary of a conductor or insulator do not obey this equation at all.

§78　Plane polarised waves in a dielectric medium

We next discuss plane waves in a uniform non-conducting medium, and show that they are of transverse type, $\mathbf{E}$ and $\mathbf{H}$ being perpendicular to the direction of propagation. Let us consider plane waves travelling with velocity V in a direction l, m, n. Then $\mathbf{E}$ and $\mathbf{H}$ must be functions of a new variable

$$u \equiv lx + my + nz - Vt. \tag{20}$$

When we say that a vector such as $\mathbf{E}$ is a function of u, we mean that each of its three components separately is a function of u, though the three functions will not in general be the same. Consider the fourth equation of (11). Its x component (see (3′)) is

$$\frac{\partial H_z}{\partial y} - \frac{\partial H_y}{\partial z} = \varepsilon_0 \varkappa_e \frac{\partial E_x}{\partial t}.$$

If dashes denote differentiation with respect to u, this is

$$mH'_z - nH'_y = -\varepsilon_0 \varkappa_e V E'_x.$$

Integrating with respect to u, this becomes

$$mH_z - nH_y = -\varepsilon_0 \varkappa_e V E_x,$$

in which we have put the constant of integration equal to zero, since we are concerned with fluctuating fields whose mean value is zero. There are two similar equations to the above, for E_y and E_z, and we may write them as one vector equation. If we let $\mathbf{n}$ denote the unit vector in the direction of propagation, so that $\mathbf{n} = (l, m, n)$ with $l^2 + m^2 + n^2 = 1$, we have

$$\mathbf{n} \times \mathbf{H} = -\varepsilon_0 \varkappa_e V \mathbf{E}. \tag{21}$$

Exactly similar treatment is possible for the third equation of (11); we get

$$\mathbf{n} \times \mathbf{E} = \mu_0 \varkappa_m V \mathbf{H}. \tag{22}$$

Equation (21) shows that $\mathbf{E}$ is perpendicular to $\mathbf{n}$ and $\mathbf{H}$, and (22) shows that $\mathbf{H}$ is perpendicular to $\mathbf{n}$ and $\mathbf{E}$. In other words, both $\mathbf{E}$ and $\mathbf{H}$ are perpendicular to the direction of propagation, so that the waves are **transverse**, and in addition, $\mathbf{E}$ and $\mathbf{H}$ are themselves perpendicular; $\mathbf{E}$, $\mathbf{H}$ and $\mathbf{n}$ forming a right-handed set of axes. If we eliminate $\mathbf{H}$ from (21) and (22) and use the fact that

$$\mathbf{n} \times [\mathbf{n} \times \mathbf{E}] = (\mathbf{n} \cdot \mathbf{E})\mathbf{n} - (\mathbf{n} \cdot \mathbf{n})\mathbf{E} = -\mathbf{E},$$

since $\mathbf{n}$ is perpendicular to $\mathbf{E}$ and $\mathbf{n}$ is a unit vector, we discover that $V^2 = c^2 / \varkappa_e \varkappa_m$, showing again that the velocity of these waves is indeed $c / \sqrt{(\varkappa_e \varkappa_m)}$.

It is worth while writing down the particular cases of (21) and (22) that correspond to plane harmonic waves in the direction of the z axis, and with the **E** vector in the x or y directions. The solutions are

$$E_x = 0 \qquad\qquad H_x = -\sqrt{(\varepsilon_0 \varkappa_e / \mu_0 \varkappa_m)} a \; e^{ip(t-z/V)}$$

$$E_y = a \; e^{ip(t-z/V)} \qquad H_y = 0 \tag{23}$$

$$E_z = 0 \qquad\qquad H_z = 0$$

$$E_x = b \; e^{ip(t-z/V)} \qquad H_x = 0$$

$$E_y = 0 \qquad\qquad H_y = +\sqrt{(\varepsilon_0 \varkappa_e / \mu_0 \varkappa_m)} b \; e^{ip(t-z/V)} \tag{24}$$

$$E_z = 0 \qquad\qquad H_z = 0.$$

In accordance with §10, a and b may be complex, the arguments giving the two phases. It is the general convention to call the plane containing **H** and **n** the **plane of polarisation**, though not all books conform to this nomenclature. Thus (23) is a wave polarised in the (x, z) plane, and (24) a wave polarised in the (y, z) plane. By the principle of superposition (§6) we may superpose solutions of types (23) and (24). If the two phases are different, we obtain **elliptically polarised light**, in which the end-point of the vector **E** describes an ellipse in the (x, y) plane. If the phases are the same, we obtain **plane polarised light**, polarised in the plane $y/x = -b/a$. If the phases differ by $\pi/2$, and the amplitudes are equal, we obtain **circularly polarised light**, which, in real form, may be written

$$E_x = a \cos p(t - z/V) \qquad H_x = -\sqrt{(\varepsilon_0 \varkappa_e / \mu_0 \varkappa_m)} a \sin p(t - z/V)$$

$$E_y = a \sin p(t - z/V) \qquad H_y = +\sqrt{(\varepsilon_0 \varkappa_e / \mu_0 \varkappa_m)} a \cos p(t - z/V)$$

$$E_z = 0 \qquad\qquad H_z = 0 \tag{25}$$

The end-points of the vectors **E** and **H** each describe circles in the (x, y) plane.

In all three cases (23)–(25), when we are dealing with free space ($\varkappa_e = \varkappa_m = 1$) the magnitudes of **E** and **H** are equal.

§79   Rate of transmission of energy in plane waves

By the use of (10) we can easily write down the rate at which energy is transmitted in these waves. Thus, with (25) the Poynting Vector is simply

$$\left(0, 0, a^2 \left(\frac{\varepsilon_0 \varkappa_e}{\mu_0 \varkappa_m} \right)^{1/2} \right).$$

This vector is in the direction of the positive z axis, showing that energy is propagated with the waves. According to (9), the total energy per unit volume is

$$\tfrac{1}{2}(\varepsilon_0 \varkappa_e \mathbf{E}^2 + \mu_0 \varkappa_m \mathbf{H}^2) = \varepsilon_0 \varkappa_e a^2.$$

From these two expressions we can deduce the velocity with which the energy flows; for this velocity is merely the ratio of the total flow across unit area in unit time divided by the energy per unit volume. This is $c/\sqrt(\varkappa_e \varkappa_m)$ so that the energy flows with the same velocity as the wave. This does not hold with all types of wave motion; an exception has already occurred in liquids (§59).

When we calculate the Poynting Vector for the waves (23) and (24), we must remember that $\mathbf{E} \times \mathbf{H}$ is not a linear function and consequently (see §10) we must choose either the real or the imaginary parts of $\mathbf{E}$ and $\mathbf{H}$. Taking, for example, the real part of (23), the Poynting Vector lies in the z direction, with magnitude

$$\left(\frac{\varepsilon_0 \varkappa_e}{\mu_0 \varkappa_m}\right)^{1/2} a^2 \cos^2 p\left(t - \frac{z}{V}\right).$$

This is a fluctuating quantity whose mean value with respect to the time is

$$\tfrac{1}{2}a^2(\varepsilon_0 \varkappa_e / \mu_0 \varkappa_m)^{1/2}.$$

The energy density, from (9), is $\varepsilon_0 \varkappa_e a^2 \cos^2 p(t - z/V)$, with a corresponding mean value $\tfrac{1}{2}\varepsilon_0 \varkappa_e a^2$. Once again the velocity of transmission of energy is $\tfrac{1}{2}a^2(\varepsilon_0 \varkappa_e / \mu_0 \varkappa_m)^{1/2} \div \tfrac{1}{2}\varepsilon_0 \varkappa_e = c/\sqrt(\varkappa_e \varkappa_m)$, which is the same as the wave velocity.

§80   Reflection and refraction of light waves

We shall next discuss the reflection and refraction of plane harmonic light waves. This reflection will be supposed to take place at a plane surface separating two non-conducting dielectric media whose refractive indices are K_1 and K_2. Since we may put $\varkappa_{1m} = \varkappa_{2m} = 1$, the velocities in the two media are c/K_1, c/K_2. In Fig. 26 let Oz be the direction of the common normal to the two media, and let AO, OB, OC be the directions of the incident, reflected and refracted (or transmitted) waves. We have not yet shown that these all lie in a plane; let us suppose that they make angles θ, $\pi - \theta'$ and ϕ with the z axis, OA being in the plane of the paper, and let us take the plane of incidence (i.e. the plane containing OA and Oz) to be the (x, z)-plane. The y axis is then perpendicular to the plane of the paper.

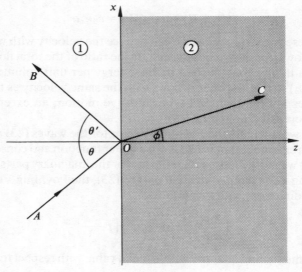

Fig. 26

Since the angle of incidence is θ, then as in (20), each of the three components of **E** and **H** will be proportional to

$$\exp\{ip[ct - K_1(x \sin\theta + z \cos\theta)]\}.$$

Let the reflected and transmitted rays move in directions with direction cosines (l_1, m_1, n_1) and (l_2, m_2, n_2), respectively, so that $n_1 = -\cos\theta'$, $n_2 = \cos\phi$. Then the corresponding components of **E** and **H** for these rays will be proportional to

$$\exp\{ip[ct - K_1(l_1x + m_1y + n_1z)]\}$$

and

$$\exp\{ip[ct - K_2(l_2x + m_2y + n_2z)]\}.$$

Thus, considering the E_x components, we may write the incident, reflected and transmitted values

$$A \exp\{ip[ct - K_1(x \sin\theta + z \cos\theta)]\},$$

$$A_1 \exp\{ip[ct - K_1(l_1x + m_1y + n_1z)]\}$$

and

$$A_2 \exp\{ip[ct - K_2(l_2x + m_2y + n_2z)]\}.$$

These functions all satisfy the standard wave equation and they have the same frequency, a condition which is necessary from the very nature of the problem.

We shall first show that the reflected and transmitted waves lie in the plane of incidence. This follows from the boundary conditions (8) that E_x must be continuous on the plane $z = 0$. That is, for all x, y, t,

$$A \exp\{ip(ct - K_1x \sin \theta)\} + A_1 \exp\{ip[ct - K_1(l_1x + m_1y)]\}$$
$$= A_2 \exp\{ip[ct - K_2(l_2x + m_2y)]\}.$$

This identity is only possible if the indices of all three terms are identical, so that we conclude

$$ct - K_1x \sin \theta \equiv ct - K_1(l_1x + m_1y) \equiv ct - K_2(l_2x + m_2y).$$

Thus

$$K_1 \sin \theta = K_1l_1 = K_2l_2,$$
$$0 = K_1m_1 = K_2m_2.$$

The second of these relations shows that $m_1 = m_2 = 0$, so that the reflected and transmitted rays OB, OC lie in the plane of incidence xOz. The first relation shows that $l_1 = \sin \theta$, so that the angle of reflection θ' is equal to the angle of incidence θ, and also

$$K_1 \sin \theta = K_2 \sin \phi. \tag{26}$$

This well-known relationship between the angles of incidence and refraction is known as **Snell's law**.

Our discussion so far has merely concerned itself with directions, and we must now pass to the amplitudes of the waves. There are two main cases to consider, according as the incident light is polarised in the plane of incidence, or perpendicular to it.

Incident light polarised in the plane of incidence

The incident ray AO has its magnetic vector in the (x, z) plane, directed perpendicular to AO. To express this vector in terms of x, y, z it is convenient to use intermediary axes ξ, η, ζ through O indicated in Fig. 27, where the directions of ξ and ζ are as shown. Here η coincides with the y axis which is perpendicular to the plane of the paper, while ζ

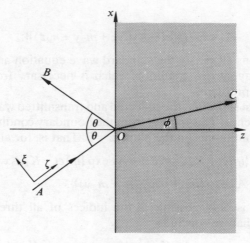

Fig. 27

is in the direction of propagation, and ξ is in the plane of incidence. Referred to these new axes, **H** lies entirely in the ξ direction, and **E** in the η direction. We may use (23) and write

$$E_\xi = E_\zeta = 0, \qquad E_\eta = a_1 \exp\{ip(ct - K_1\zeta)\}$$
$$H_\eta = H_\zeta = 0, \qquad H_\xi = -K_1 a_1 \exp\{ip(ct - K_1\zeta)\}.$$

Now $\zeta = x \sin\theta + z \cos\theta$, and so it follows that:

incident wave

$$E_x = 0, \qquad H_x = -K_1 a_1 \cos\theta \exp\{ip[ct - K_1(x \sin\theta + z \cos\theta)]\}.$$
$$E_y = a_1 \exp\{ip[ct - K_1(x \sin\theta + z \cos\theta)]\}, \qquad H_y = 0,$$
$$E_z = 0, \qquad H_z = K_1 a_1 \sin\theta \exp\{ip[ct - K_1(x \sin\theta + z \cos\theta)]\}.$$

Similar analysis for the reflected and refracted waves in which we replace θ by $\pi - \theta$ and ϕ in turn, enables us to write

reflected wave

$$E_x = 0, \qquad H_x = K_1 b_1 \cos\theta \exp\{ip[ct - K_1(x \sin\theta - z \cos\theta)]\},$$
$$E_y = b_1 \exp\{ip[ct - K_1(x \sin\theta - z \cos\theta)]\}, \qquad H_y = 0,$$
$$E_z = 0, \qquad H_z = K_1 b_1 \sin\theta \exp\{ip[ct - K_1(x \sin\theta - z \cos\theta)]\}:$$

refracted wave

$$E_x = 0, \qquad H_x = -K_2 a_2 \cos \phi \exp \{ip[ct - K_2(x \sin \phi + z \cos \phi)]\}.$$

$$E_y = a_2 \exp \{ip[ct - K_2(x \sin \phi + z \cos \phi)]\}, \qquad H_y = 0,$$

$$E_z = 0, \qquad H_z = K_2 a_2 \sin \phi \exp \{ip[ct - K_2(x \sin \phi + z \cos \phi)]\}.$$

We may write the boundary conditions in the form that E_x, E_y, KE_z, H_x, H_y and H_z are continuous at $z = 0$. These six conditions reduce to two independent relations, which we may take to be those due to E_y and H_x:

$$a_1 + b_1 = a_2,$$

$$-K_1 a_1 \cos \theta + K_1 b_1 \cos \theta = -K_2 a_2 \cos \phi.$$

Thus

$$\frac{a_1}{K_1 \cos \theta + K_2 \cos \phi} = \frac{b_1}{K_1 \cos \theta - K_2 \cos \phi} = \frac{a_2}{2K_1 \cos \theta}.$$

Using Snell's law (26) in the form $K_1 : K_2 = \sin \phi : \sin \theta$, this gives

$$\frac{a_1}{\sin (\theta + \phi)} = \frac{b_1}{-\sin (\theta - \phi)} = \frac{a_2}{2 \sin \phi \cos \theta}. \tag{27}$$

Equation (27) gives the ratio of the reflected and refracted amplitudes. If medium 2 is denser than medium 1, $K_2 > K_1$, so that $\theta > \phi$, and thus b_1/a_1 is negative; so there is a phase change of π in the electric field when reflection takes place in the lighter medium. There is no phase change on reflection in a denser medium, nor in the refracted wave. The same conclusion is true for the magnetic field $\mathbf{H}$, though it must be remembered here and later in this chapter that the positive directions for $\mathbf{E}$ and $\mathbf{H}$ are defined by $\xi \eta \zeta$ in Fig. 27 and their counterparts relative to OB, OC, the η direction being along the y axis throughout.

Incident light polarised perpendicular to the plane of incidence

A similar discussion can be given when the incident light is polarised perpendicular to the plane of incidence; in this case the roles of $\mathbf{E}$ and $\mathbf{H}$ are practically interchanged, H_y for example being the only non-vanishing component of $\mathbf{H}$. It is not necessary to repeat the analysis in full. With the same notation for the amplitudes of the incident, reflected and refracted waves, we have

$$\frac{a_1}{\sin 2\theta + \sin 2\phi} = \frac{b_1}{\sin 2\theta - \sin 2\phi} = \frac{a_2}{4\cos\theta\sin\phi}. \qquad (28)$$

It follows from (28) that the reflected ray vanishes if $\sin 2\theta = \sin 2\phi$. Since $\theta \neq \phi$, this implies that $\theta + \phi = \pi/2$, and then Snell's law gives

$$K_1 \sin\theta = K_2 \sin\phi = K_2 \cos\theta,$$

so

$$\tan\theta = K_2/K_1 = \sqrt{(\kappa_{2e}/\kappa_{1e})}. \qquad (29)$$

With this angle of incidence, known as **Brewster's angle**, there is no reflected ray.

There will be a phase change of π in **E** on reflection when $\sin 2\theta < \sin 2\phi$. If reflection takes place in the lighter medium so that $K_1 < K_2$, this holds for θ greater than Brewster's angle; but if $K_1 > K_2$, it holds for θ less than Brewster's angle. In all other cases there is no phase change on reflection.

In general, of course, the incident light is composed of waves polarised in all possible directions. Equations (27) and (28) show that if the original amplitudes in the two main directions are equal, the reflected amplitudes will not be equal, so that the light becomes partly polarised on reflection. When the angle of incidence is given by (29) it is completely polarised on reflection. This angle is therefore sometimes known as the **polarising angle**.

§81   Internal reflection

An interesting possibility arises in the discussion of **§80**, which gives rise to the phenomenon known as **total** or **internal reflection.** It arises when reflection takes place in the denser medium so that $\phi > \theta$. If we suppose θ to be steadily increased from zero, then ϕ also increases and when $\sin\theta = K_2/K_1$, $\phi = \pi/2$. If θ is increased beyond this critical value, ϕ is imaginary. There is nothing to disturb us in this fact provided that we interpret the analysis of **§80** correctly, for we never had occasion to suppose that the coefficients were real. We can easily make the necessary adjustment in this case. Take for simplicity the case of incident light polarised in the plane of incidence. Then the incident and reflected waves are just as in our previous calculations. The refracted wave has the same form also, but in the exponential term, $K_2 \sin \phi = K_1 \sin \theta$, and is therefore real, whereas

$$K_2 \cos \phi = \surd(K_2^2 - K_2^2 \sin^2 \phi) = \surd(K_2^2 - K_1^2 \sin^2 \theta),$$

and is imaginary, since we are supposing that internal reflection is taking place and therefore $K_1 \sin \theta > K_2$. We may therefore write $K_2 \cos \phi = \pm iq$, where q is real. Thus the refracted wave has the form

$$E_y = a_2 \exp \{ip[ct - K_1 \sin \theta x \pm iqz]\}$$

$$= a_2 \exp (\pm pqz) \,.\, \exp \{ip[ct - K_1 \sin \theta x]\}.$$

For reasons of finiteness at infinity, we have to choose the negative sign, so that it appears that the wave is attenuated as it proceeds into the less dense medium. For normal light waves it turns out that the penetration is only a few wavelengths, and this justifies the title of total reflection. The decay factor is

$$\exp (-pqz) = \exp \{-p\surd(K_1^2 \sin^2 \theta - K_2^2)z\}.$$

This factor increases with the frequency so that light of great frequency hardly penetrates at all. In actual physical problems, the refractive index does not change from K_1 to K_2 abruptly, as we have imagined; however, Drude has shown that if we suppose that there is a thin surface layer, of thickness approximately equal to one atomic diameter, in which the change takes place smoothly, the results of this and the preceding paragraphs are hardly affected.

§82 Partially conducting media, plane waves

In our previous calculations we have assumed that the medium was nonconducting, so that we could put $\sigma = 0$. When we remove this restriction, keeping always to homogeneous media, equations (1)–(7) give us

$$\text{div } \mathbf{E} = 0,$$

$$\text{div } \mathbf{H} = 0,$$

$$\text{curl } \mathbf{H} = \sigma \mathbf{E} + \varepsilon_0 \varkappa_e \frac{\partial \mathbf{E}}{\partial t},$$

$$\text{curl } \mathbf{E} = -\mu_0 \varkappa_m \frac{\partial \mathbf{H}}{\partial t}.$$

Now we have the identity

$$\text{curl curl } \mathbf{E} = \text{grad div } \mathbf{E} - \nabla^2 \mathbf{E} = -\nabla^2 \mathbf{E},$$

so that

$$\nabla^2 \mathbf{E} = \mu_0 \varkappa_m \text{ curl } \frac{\partial \mathbf{H}}{\partial t} = \mu_0 \varkappa_m \frac{\partial}{\partial t} \text{ curl } \mathbf{H} = \sigma \mu_0 \varkappa_m \frac{\partial \mathbf{E}}{\partial t} + \frac{\varkappa_e \varkappa_m}{c^2} \frac{\partial^2 \mathbf{E}}{\partial t^2},$$

and so

$$\nabla^2 \mathbf{E} = \frac{\varkappa_e \varkappa_m}{c^2} \frac{\partial^2 \mathbf{E}}{\partial t^2} + \sigma \mu_0 \varkappa_m \frac{\partial \mathbf{E}}{\partial t}. \tag{30}$$

A similar equation holds for $\mathbf{H}$. Equation (30) will be recognised as the equation of telegraphy (see §9). The first term on the right-hand side may be called the displacement term, since it arises from the displacement current $\dfrac{\partial \mathbf{D}}{\partial t}$ and the second is the conduction term, since it arises from the conduction current $\mathbf{j}$. If we are dealing with waves whose frequency is $p/2\pi$, $\mathbf{E}$ will be proportional to e^{ipt}; the ratio of the first to the second term is therefore of the order $\varepsilon_0 \varkappa_e p / \sigma$. This means that if $p \gg \varepsilon_e / \sigma$, only the displacement term matters (this is the case of light waves in a non-conducting dielectric); but if $p \ll \varepsilon_0 \varkappa_e / \sigma$, only the conduction term matters (this is the case of long waves in a good metallic conductor). In the intermediate region both terms must be retained. With most metals, if $p < 10^7$ per second we can neglect the first term, and if $p > 10^{15}$ per second we can neglect the second term.

Let us discuss the solutions of (30) which apply to plane harmonic waves propagated in the z direction, such that only E_x and H_y are non-vanishing (as in (24)). We may suppose that each of these components is proportional to

$$\exp\{ip(t - qz)\}. \tag{31}$$

where $p/2\pi$ is the frequency and q is still to be determined. This expression satisfies the equation (30) if

$$q^2 = \frac{\varkappa_e \varkappa_m}{c^2} \left\{ 1 - \frac{\sigma}{\varepsilon_0 \varkappa_e p} i \right\}, \tag{32}$$

so that q is therefore complex, and we may write it

$$q = \alpha - i\beta,$$

where

$$\alpha^2 = \frac{\varkappa_e \varkappa_m}{2c^2} \left[\left\{ 1 + \left(\frac{\sigma}{\varepsilon_0 \varkappa_e p} \right)^2 \right\}^{1/2} + 1 \right]$$

$$\beta^2 = \frac{\varkappa_e \varkappa_m}{2c^2} \left[\left\{ 1 + \left(\frac{\sigma}{\varepsilon_0 \varkappa_e p} \right)^2 \right\}^{1/2} - 1 \right]. \tag{33}$$

The "velocity" of (31) is $1/q$; but we have seen in §80 that in a medium of refractive index K the velocity is c/K. So the effective refractive index is cq which is complex. Complex refractive indices occur quite frequently and are associated with absorption of the waves; for, combining (31) and (33) we have the result that E_x and H_y are proportional to

$$\exp\{-p\beta z\} \cdot \exp\{ip(t - \alpha z)\}. \tag{34}$$

This shows that a plane wave cannot be propagated in such a medium without absorption. The decay factor may be written e^{-kz} where $k = p\beta$. k is called the **absorption coefficient**. In the case where $\sigma/\varepsilon_0 \varkappa_e p$ is small compared with unity (the case of light waves in most non-conducting dielectrics), k is approximately equal to $(\sigma/2c\varepsilon_0)\sqrt{(\varkappa_m/\varkappa_e)}$. Now the wavelength in (34) is $\lambda = 2\pi/\alpha p$, so that in one wavelength the amplitude decays by a factor $e^{-k\lambda}$, approximately $\exp(-\pi\sigma/\varepsilon_0 \varkappa_e p)$. As we are making the assumption that $\sigma/\varepsilon_0 \varkappa_e p$ is small, the decay is gradual, and can only be noticed after many wavelengths. The distance travelled before the amplitude is reduced to $1/e$ times its original value is $1/k$, which is of the same order as σ.

The velocity of propagation of (34) is $1/\alpha$, and thus varies with the frequency. With our usual approximation that $\sigma/\varepsilon_0 \varkappa_e p$ is small, this velocity is

$$\frac{c}{\sqrt{\varkappa_e \varkappa_m}} \left\{ 1 - \left(\frac{\sigma}{2\varepsilon_0 \varkappa_e p} \right)^2 \right\}. \tag{35}$$

We can show that in waves of this character **E** and **H** are out of phase with each other. For if in accordance with (31), we write

$$E_x = a \exp\{ip(t - qz)\},$$
$$H_y = b \exp\{ip(t - qz)\},$$

then the y-component of the vector relation

$$\text{curl } \mathbf{E} = -\mu_0 \varkappa_m \frac{\partial \mathbf{H}}{\partial t},$$

gives us the connection between a and b. It is

$$\frac{\partial E_x}{\partial z} = -\mu_0 \varkappa_m \frac{\partial H_y}{\partial t},$$

which is equivalent to

$$qa = \mu_0 \varkappa_m b. \tag{36}$$

Now q is complex and hence there is a phase difference between E_x and H_y equal to the argument of q. This is $\tan^{-1}(\beta/\alpha)$, and with the same approximation as in (35), this is just $\tan^{-1}(\sigma/2\varepsilon_0 \varkappa_e p)$, which is effectively $\sigma/2\varepsilon_0 \varkappa_e p$.

§83   Reflection from a metal

It is interesting to discuss in more detail the case in which the conductivity is so great that we may completely neglect the displacement term in (30). Let us consider the case of a beam of light falling normally on an infinite metallic conductor bounded by the plane $z = 0$. Let us suppose also, as in Fig. 28, that the incident waves come from

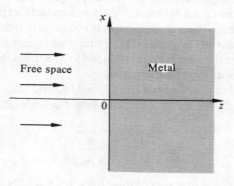

Free space   Metal

Fig. 28

the negative direction of z, in free space, for which $\varkappa_e = \varkappa_m = 1$, and are polarised in the (y, z)-plane. Then, according to (24), they are defined by:

incident wave

$$E_x = a_1\{ip(t - z/c)\}, \qquad H_y = a_1 \exp\{ip(t - z/c)\}, \tag{37a}$$

reflected wave

$$E_x = b_1 \exp\{ip(t + z/c)\}, \qquad H_y = -b_1 \exp\{ip(t + z/c)\}.$$

(37b)

In the metal itself we may write, according to (31) and (36),

$$E_x = a_2 \exp\{ip(t - qz)\}, \qquad H_y = (qa_2/\mu_0 \varkappa_m) \exp\{ip(t - qz)\}. \qquad (37c)$$

These values will satisfy the equation of telegraphy (30) in which we have neglected the displacement term, if

$$q^2 = -\frac{\sigma\mu_0\varkappa_m}{p}i = -2\gamma^2 i,$$

where

$$\gamma^2 = \sigma\mu_0\varkappa_m/2p,$$

so that

$$q = \gamma(1 - i). \qquad (37d)$$

Inside the metal, $\mathbf{E}$ and $\mathbf{H}$ have a $\pi/4$ phase difference, since, as we have shown in (36), this phase difference is merely the argument of q.

The boundary conditions are that E_x and H_y are continuous at $z = 0$. This gives two equations

$$a_1 + b_1 = a_2,$$

$$a_1 - b_1 = qa_2/\mu_0\varkappa_m.$$

Hence

$$\frac{a_1}{1 + q/\mu_0\varkappa_m} = \frac{b_1}{1 - q/\mu_0\varkappa_m} = \frac{a_2}{2}. \qquad (38)$$

Since q is complex, all three electric vectors have phase differences. The ratio R of reflected to incident energy is $|b_1/a_1|^2$, which reduces to

$$\frac{(\gamma - \mu_0\varkappa_m)^2 + \gamma^2}{(\gamma + \mu_0\varkappa_m)^2 + \gamma^2}.$$

In the case of non-ferromagnetic metals, γ is much larger than $\mu_0\varkappa_m$, so that approximately

$$R = 1 - \frac{2\mu_0\varkappa_m}{\gamma}.$$

This formula has been checked excellently by experiments for wavelengths in the region of 10^{-5} cms.

It is an easy matter to generalise these results to apply to the case when we include both the displacement and conduction terms in (30).

We can use (38) to calculate the loss of energy in the metal. If we consider unit area of the surface of the metal, the rate of arrival of energy is given by the Poynting Vector. This is $\frac{1}{2}|a_1|^2$. Similarly the rate of reflection of energy is $\frac{1}{2}|b_1|^2$. So the rate of dissipation is $\frac{1}{2}\{|a_1|^2 - |b_1|^2\}$. This must be the same as the Joule heat loss. In our units, this loss is $\sigma \mathbf{E}^2$ per unit volume per unit time. If we take the mean value of E_x^2 in the metal, it is a straightforward matter to show that $\int_0^\infty \sigma E_x^2 \, dz$ is indeed exactly equal to this rate of dissipation.

§84   Radiation pressure

When the radiation falls on the metal of **§83**, it exerts a pressure. We may calculate this, if we use the experimental law that when a current $\mathbf{j}$ is in the presence of a magnetic field $\mathbf{H}$ there is a force $\mu_0 \varkappa_m \mathbf{j} \times \mathbf{H}$ acting on it. In our problem, there is, in the metal, an alternating field $\mathbf{E}$, and a corresponding current $\sigma \mathbf{E}$. The force on the current is therefore $\mu_0 \varkappa_m \sigma \mathbf{E} \times \mathbf{H}$, and this force, being perpendicular to $\mathbf{E}$ and $\mathbf{H}$, lies in the z direction. The force on the charges that compose the current is transmitted by them to the metal as a whole. Now both $\mathbf{E}$ and $\mathbf{H}$ are proportional to $e^{-p\gamma z}$ (see equations (37c,d)) so that the force falls off according to the relation $e^{-2p\gamma z}$. To calculate the total force on a unit area of the metal surface, we must integrate $\mu_0 \varkappa_m \sigma \mathbf{E} \times \mathbf{H}$ from $z = 0$ to $z = \infty$. $\mathbf{E} \times \mathbf{H}$ is a fluctuating quantity, and so we shall have to take its mean value with respect to the time. The pressure is then

$$(\mu_0 \varkappa_m \sigma)(1/\mu_0 \varkappa_m)|a_2|^2 \int_0^\infty \tfrac{1}{2}\gamma \, e^{-2p\gamma z} \, dz,$$

or,

$$(\sigma/4p)|a_2|^2.$$

Using (38) this may be expressed in the form

$$(\sigma \mu_0^2 \varkappa_m^2 / p)|a_1|^2 \{(\gamma + \mu_0 \varkappa_m)^2 + \gamma^2\}^{-1}.$$

§85   Skin effect

There is another application of the theory of **§83** which is important. Suppose that we have a straight wire of circular section, and a rapidly alternating e.m.f. is applied at its two ends. We have seen in **§83** that with an infinite sheet of metal the current falls off as we penetrate the metal according to the law $e^{-p\gamma z}$. If $p\gamma$ is small, there is little attenuation as we go down a distance equal to the radius of the wire, and clearly the current will be almost constant for all parts of any section (see, however, question (12) in §87). But if $p\gamma$ is large, then the current will be carried mainly near the surface of the wire, and it will not make a great deal of difference whether or not the metal is infinite in extent, as we suppose in §83, or whether it has a cross-section in the form of a circle; in this case the current density falls off approximately according to the law $e^{-p\gamma r}$ as we go down a distance r from the surface. This phenomenon is known as the **skin effect**; it is more pronounced at very high frequencies.

We could, of course, solve the problem of the wire quite rigorously, using cylindrical polar coordinates. The formulae are rather complicated, but the result is essentially the same.

§86 Propagation in waveguides

The last topic we examine in this chapter is the propagation of electromagnetic waves inside long straight hollow metal ducts. These are called **waveguides**, and as the duct will be assumed to be empty, the waves will propagate in a region in which $\rho = \sigma = 0$ and $\varkappa_e = \varkappa_m = 1$. From (39) the electric vector **E** will then satisfy the vector wave equation

$$\nabla^2 \mathbf{E} = \frac{1}{c^2} \frac{\partial^2 \mathbf{E}}{\partial t^2}, \tag{39}$$

while from the first Maxwell equation (1) we must also require of **E** that

$$\text{div } \mathbf{E} = 0. \tag{40}$$

Our task will thus be to seek a solution of (39) that satisfies (40), and which also obeys appropriate boundary conditions for **E** on the walls of the duct which we will assume to be perfect conductors. We shall take the z axis to lie along the duct and to have associated with it the unit vector **k**. The duct itself will be supposed to be of constant cross section.

The task of solving this problem will be much simplified if we notice that the vector

$$\mathbf{E} = \text{curl } (u\mathbf{k}) \tag{41}$$

will satisfy both (39) and (40), provided only that the scalar u satisfies the scalar wave equation

$$\nabla^2 u = \frac{1}{c^2} \frac{\partial^2 u}{\partial t^2}. \tag{42}$$

This result follows because

$$\nabla^2 [\text{curl } (u\mathbf{k})] = \text{curl } [\nabla^2 (u\mathbf{k})],$$

$$\frac{\partial^2}{\partial t^2} [\text{curl } (u\mathbf{k})] = \text{curl } \left[\frac{\partial^2}{\partial t^2} (u\mathbf{k}) \right]$$

and

$$\text{div curl } (u\mathbf{k}) \equiv 0.$$

If we consider harmonic waves with frequency $p/2\pi$ that propagate in the z direction with wave number $k/2\pi$ we may set

$$u(\mathbf{r}, t) = \phi(x, y) \exp \{i(pt - kz)\}, \tag{43}$$

in which $\mathbf{r}$ is the general position vector in the duct. It is then a consequence of (41) and (43) that

$$E_x = \frac{\partial \phi}{\partial y} \exp \{i(pt - kz)\}, \qquad E_y = -\frac{\partial \phi}{\partial x} \exp \{i(pt - kz)\}, \qquad E_z = 0. \tag{44}$$

The harmonic time behaviour assumed for $\mathbf{E}$ will also apply to $\mathbf{B}$, so that once this time dependence has been taken into account Maxwell's equations (3) and (4) can be put in the form

$$\hat{\mathbf{E}} = -\frac{ic^2}{p} \text{curl } \hat{\mathbf{B}} \tag{45}$$

$$\hat{\mathbf{B}} = \frac{i}{p} \text{curl } \hat{\mathbf{E}}, \tag{46}$$

where we have separated out the time dependence by writing

$$\mathbf{E}(\mathbf{r}, t) = \hat{\mathbf{E}}(\mathbf{r}) \exp (ipt) \quad \text{and} \quad \mathbf{B}(\mathbf{r}, t) = \hat{\mathbf{B}}(\mathbf{r}) \exp (ipt). \tag{47}$$

Comparing (44) with the first expression in (47) then shows that $\hat{\mathbf{E}}(\mathbf{r})$ has the components

$$E_x = \frac{\partial \phi}{\partial y} \exp (-ikz), \qquad E_y = -\frac{\partial \phi}{\partial x} \exp (-ikz), \qquad E_z = 0. \tag{48}$$

When used in (46) this result leads to the following components for $\hat{\mathbf{B}}(\mathbf{r})$

$$\hat{B}_x = -\frac{k}{p}\hat{E}_y, \qquad \hat{B}_y = \frac{k}{p}\hat{E}_x, \qquad \hat{B}_z = -\frac{i}{p}(\nabla^2\phi)\exp(-ikz). \quad (49)$$

Inspection of (48) and (49) shows that $\hat{\mathbf{E}}$ and $\hat{\mathbf{B}}$ are orthogonal, because $\hat{\mathbf{E}} \cdot \hat{\mathbf{B}} = 0$. Although $\hat{\mathbf{E}}$ lies purely in the (x, y)-plane, and so is transverse to the direction of propagation $\mathbf{k}$, the vector $\hat{\mathbf{B}}$ will only be transverse if $\nabla^2\phi = 0$. To emphasise the transverse nature of $\hat{\mathbf{E}}$, a wave of this type in which $\hat{B}_z \neq 0$ will be called a **transverse electric wave**, which is usually abbreviated to a TE wave. If, however, $\nabla^2\phi = 0$, so that $\hat{\mathbf{B}}$ is also purely transverse to the direction of propagation, then the wave will be called a **transverse electromagnetic wave**, or a TEM wave.

To proceed further we now observe that the vector

$$\mathbf{V} = \text{curl curl}(u\mathbf{k}) \quad (50)$$

is also a solution of (39) and (40). Then, because of the relationship that exists between the time independent parts $\hat{\mathbf{E}}$ and $\hat{\mathbf{V}}$ of (41) and (50), it can be seen by inspection of (45) and (46) that when $\mathbf{E}$ is given by (41), $\mathbf{B}$ will be given by (50), and conversely. So, analogously to the previous case, we find that when $\mathbf{B}$ is given by (50),

$$\hat{B}_x = \frac{\partial\phi}{\partial y}\exp(-ikz), \qquad \hat{B}_y = -\frac{\partial\phi}{\partial x}\exp(-ikz), \qquad \hat{B}_z = 0 \quad (51)$$

and

$$\hat{E}_x = \frac{kc^2}{p}\hat{B}_y, \qquad \hat{E}_y = -\frac{kc^2}{p}\hat{B}_x, \qquad \hat{E}_z = \frac{ic^2}{p}(\nabla^2\phi)\exp(-ikz). \quad (52)$$

Here again the vectors $\hat{\mathbf{E}}$ and $\hat{\mathbf{B}}$ are orthogonal since $\hat{\mathbf{E}} \cdot \hat{\mathbf{B}} = 0$, but this time it is the $\hat{\mathbf{B}}$ that is purely transverse and $\hat{\mathbf{E}}$ that will only be transverse if $\nabla^2\phi = 0$. For this reason, when $\hat{E}_z \neq 0$, waves of this type are called **transverse magnetic waves**, or TM waves.

We have thus identified three essentially different modes of propagation in a waveguide according as the waves in question are of the TE, TEM or TM types. The detailed nature of the wave propagation in each case will follow once ϕ has been found from the boundary conditions that have still to be specified.

To solve for ϕ we now substitute (43) into (42) to obtain

$$\nabla^2\phi + \varkappa^2\phi = 0, \quad (53)$$

where

$$x^2 = \frac{p^2}{c^2} - k^2. \tag{54}$$

Equation (53) is called the **Helmholtz equation** and it is this equation that must be solved if ϕ, and hence $\mathbf{E}$ and $\mathbf{B}$, are to be determined. The boundary conditions on ϕ will ensure that only certain values of the constant x will be permitted. These will be the eigenvalues of the Helmholtz equation, while the corresponding solutions ϕ will be the associated eigenfunctions. Each such eigenfunction ϕ will characterise a different fundamental form of wave propagation that is associated with the TE, TEM or TM wave that is being propagated.

In passing, we notice from (54) that the wave number $k/2\pi$ will only be real if $p^2/c^2 > x^2$. If this condition on the frequency $p/2\pi$ is not satisfied k will become imaginary and the waves will attenuate exponentially with z. Thus, for any particular x, the value $p_c = xc$ will define a **cutoff frequency** for unattenuated wave propagation.

To conclude, we shall consider the propagation of the TE mode in a waveguide of rectangular cross section of sides a and b as shown in Fig. 29.

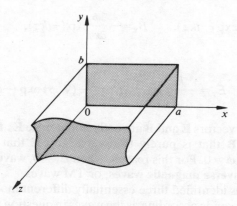

Fig. 29

Using the method of separation of variables by writing

$$\phi(x, y) = X(x) Y(y), \tag{55}$$

and proceeding as in Chapter 1, §7, we see that after division by XY, (53) becomes

$$\frac{1}{X}\frac{d^2X}{dx^2} = -\frac{1}{Y}\frac{d^2Y}{dy^2} - \varkappa^2 = -\alpha^2, \tag{56}$$

where $-\alpha^2$ is the separation constant. Consequently,

$$\frac{d^2X}{dx^2} + \alpha^2 X = 0, \tag{57}$$

and

$$\frac{d^2Y}{dy^2} + (\varkappa^2 - \alpha^2)Y = 0. \tag{58}$$

To proceed further we must now introduce boundary conditions for ϕ. As the walls are perfectly conducting, the appropriate boundary condition for the TE mode will be that the tangential component of $\hat{\mathbf{E}}$ must vanish on the walls. In the simple geometry of the rectangular waveguide of Fig. 29 this is equivalent to requiring $\hat{E}_y = 0$ on $x = 0$ and $x = a$ and $\hat{E}_x = 0$ on $y = 0$ and $y = b$. From (48) and (55) this is seen to be equivalent to

$$\frac{\partial\phi}{\partial x} = \frac{dX}{dx} = 0 \quad \text{on } x = 0 \quad \text{and} \quad x = a \tag{59}$$

and

$$\frac{\partial\phi}{\partial y} = \frac{dY}{dy} = 0 \quad \text{on } y = 0 \quad \text{and } y = b. \tag{60}$$

Solving (57) gives

$$X = A\cos\alpha x + B\sin\alpha x, \tag{61}$$

which to satisfy boundary condition (59) requires that $B = 0$ and $\sin\alpha a = 0$, yielding

$$\alpha_m = m\pi/a \quad \text{for } m = 0, 1, 2, \ldots. \tag{62}$$

When solving (58) we now set $\alpha = \alpha_m$ and

$$\beta^2 = \varkappa^2 - \alpha_m^2 \tag{63}$$

to obtain

$$Y = C\cos\beta y + D\sin\beta y. \tag{64}$$

For Y to satisfy the boundary condition (60) we must have $D = 0$ and $\sin \beta b = 0$, showing that

$$\beta_n = n\pi/b \quad \text{for } n = 0, 1, 2, \dots. \tag{65}$$

Combining (62), (64) and (65) shows that the condition defining $\varkappa$ is

$$\varkappa^2_{mn} = \left(\frac{m\pi}{a}\right)^2 + \left(\frac{n\pi}{b}\right)^2 \quad \text{for } m, n = 0, 1, 2, \dots, \tag{66}$$

when the corresponding solution ϕ_{mn} will be

$$\phi_{mn} = K \cos \frac{m\pi x}{a} \cos \frac{n\pi y}{b} \exp(-ikz), \tag{67}$$

with K an arbitrary constant.

The components of $\mathbf{E}$ in the TE (m, n) mode of wave propagation as defined in (44) are thus

$$E_x = -\left(\frac{n\pi K}{b}\right) \cos \frac{m\pi x}{a} \sin \frac{n\pi y}{b} \exp\{i(pt - kz)\},$$

$$E_y = \left(\frac{m\pi K}{a}\right) \sin \frac{m\pi x}{a} \cos \frac{n\pi y}{b} \exp\{i(pt - kz)\},$$

$$E_z = 0.$$

The components of $\mathbf{B}$ follow directly from (46) when it is combined with the above results, and they are as follows

$$B_x = -\left(\frac{km\pi K}{pa}\right) \sin \frac{m\pi x}{a} \cos \frac{n\pi y}{b} \exp\{i(pt - kz)\},$$

$$B_y = -\left(\frac{kn\pi K}{pb}\right) \cos \frac{m\pi x}{a} \sin \frac{n\pi y}{b} \exp\{i(pt - kz)\},$$

$$B_z = \left(\frac{i\pi^2 K}{p}\right)\left(\frac{m^2}{a^2} + \frac{n^2}{b^2}\right) \cos \frac{m\pi x}{a} \cos \frac{n\pi y}{b} \exp\{i(pt - kz)\}.$$

It is possible to deduce an important result from (54) if we define the **free space wavelength** $\lambda_0 = 2\pi c/p$, the **cutoff wavelength** $\lambda_c = 2\pi/\varkappa$ and the **guide wavelength** $\lambda_g = 2\pi/k$. For then (54) can be re-written as

$$\lambda_g = \frac{\lambda_0}{[1 - (\lambda_0/\lambda_c)^2]^{1/2}}, \tag{68}$$

showing that the guide wavelength is greater than the free space wavelength. The cutoff frequency and wavelength for the waveguide depend, via $\varkappa_{mn}$, on the dimensions of the waveguide cross-section and on the mode (m, n) that is propagated. The typical frequency range of operation for waveguides is between 10^9 and 10^{11} Hz. In general, a waveguide is designed so that for any particular frequency of operation only one undamped mode will propagate.

§87 Examples

1. Prove the equations (17') and (18') in **§77**.

2. Find the value of **H** when $E_x = E_y = 0$, and $E_z = A \cos nx \cos nct$. It is given that $\mathbf{H} = \mathbf{0}$ when $t = 0$, and also $\varkappa_e = \varkappa_m = 1$, $\rho = \sigma = 0$. Show that there is no mean flux of energy in this problem.

3. Prove the equation (28) in **§80**, for reflection and refraction of light polarised perpendicular to the plane of incidence.

4. Show that the polarising angle is less than the critical angle for internal reflection. Calculate the two values if $K_1 = 6$, $K_2 = 1$.

5. Show that the reflection coefficient from glass to air at normal incidence is the same as from air to glass, but that the two phase changes are different.

6. Light falls normally on the plane face which separates two media K_1 and K_2. Show that a fraction R of the energy is reflected, and T is transmitted where

$$R = \left(\frac{K_2 - K_1}{K_2 + K_1}\right)^2, \qquad T = \frac{4K_1 K_2}{(K_2 + K_1)^2}.$$

Hence prove that if light falls normally on a slab of dielectric, bounded by two parallel faces, the total fraction of energy reflected is

$$\frac{(K_2 - K_1)^2}{K_1^2 + K_2^2},$$

and transmitted is

$$\frac{2K_1 K_2}{K_1^2 + K_2^2}.$$

It is necessary to take account of the multiple reflections that take place at each boundary.

7. Light passes normally through the two parallel faces of a piece of plate glass, for which $K = 1\cdot5$. Find the fraction of incident energy transmitted, taking account of reflection at the faces.

8. Show that when internal reflection (§81) is taking place, there is a phase change in the reflected beam. Evaluate this numerically for the case of a beam falling at an angle of $60°$ to the normal when $K_1 = 1\cdot6$, $K_2 = 1$, the light being polarised in the plane of incidence.

9. Show that if $\varkappa_m = 1$, then the reflection coefficient with metals (§83) may be written in the form $R = 1 - \sqrt{(\pi\nu\mu_0/\sigma)}$, where ν is the frequency.

10. A current flows in a straight wire whose cross section is a circle of radius a. The conduction current $\mathbf{j}$ depends only on r the radial distance from the centre of the wire, and the time t. Assuming that the displacement current can be neglected, prove that $\mathbf{H}$ is directed perpendicular to the radius vector. If $j(r, t)$ and $H(r, t)$ represent the magnitudes of $\mathbf{j}$ and $\mathbf{H}$, prove that

$$\frac{\partial}{\partial r}(Hr) = rj, \qquad \frac{\partial j}{\partial r} = \mu_0\varkappa_m\sigma\frac{\partial H}{\partial t}.$$

11. Use the results of question (10) to prove that j satisfies the differential equation

$$\frac{1}{r}\frac{\partial}{\partial r}\left(r\frac{\partial j}{\partial r}\right) = \mu_0\varkappa_m\sigma\frac{\partial j}{\partial t}.$$

By using the formula for curl in cylindrical polars twice in succession shows that $H[= H_\theta(r, t)]$ satisfies the equation

$$\frac{\partial^2 H}{\partial r^2} + \frac{1}{r}\frac{\partial H}{\partial r} - \frac{H}{r^2} = [\text{curl curl } H]_\theta = \mu_0\varkappa_m\sigma\frac{\partial H}{\partial t}.$$

Use the method of separation of variables to prove that there is a solution of the j-equation of the form $j = f(r) \exp(ipt)$ where

$$\frac{d^2f}{dr^2} + \frac{1}{r}\frac{df}{dr} - iaf = 0, \qquad a = \mu_0\varkappa_m\sigma p.$$

Hence show that f is a combination of Bessel functions of order zero and complex argument.

12. If α in question (11) is small, show that an approximate solution of the current equation is $j = A(1 + \frac{1}{4}i\alpha r^2 - \frac{1}{64}a^2r^4)\exp(ipt)$, where A is a

constant. Hence show that the total current fluctuates between $\pm J$, where, neglecting powers of α above the second $J = \pi a^2 A (1 + a^4 \alpha^2 / 384)$. Use this result to show that the heat developed in unit length of the wire in unit time is $\dfrac{J^2}{2\pi\sigma a^2} (1 + a^4 \alpha^2 / 192)$. (Questions (10), (11) and (12) are the problem of the skin effect at low frequencies.)

13. Complete the calculations leading to the results in equations (48) and (49) of §86.

14. Complete the calculations leading to the results in equations (51) and (52) of §86.

15. Prove that if $\phi(\mathbf{r})$ is a solution of the scalar Helmholtz equation

$$\nabla^2 \phi + \lambda^2 \phi = 0,$$

and $\mathbf{k}$ is a constant vector, the vectors

$$\mathbf{X} = \mathrm{curl}\,(\phi\mathbf{k}) \quad \text{and} \quad \mathbf{Y} = \frac{1}{\lambda}\,\mathrm{curl}\,\mathbf{X}$$

are independent solutions of the vector Helmholtz equation

$$\nabla^2 \mathbf{V} + \lambda^2 \mathbf{V} = \mathbf{0},$$

where $\mathbf{V}$ is a solenoidal vector (i.e. div $\mathbf{V} = 0$).
  Hence, by showing that

$$\mathrm{curl}\,(\mathbf{X} + \mathbf{Y}) = \lambda(\mathbf{X} + \mathbf{Y}),$$

deduce that $\mathbf{V}$ has the general solution

$$\mathbf{V} = \mathrm{curl}\,(u\mathbf{k}) + \frac{1}{\lambda}\,\mathrm{curl}\,\mathrm{curl}\,(u\mathbf{k}).$$

16. Show that in the TM mode for the waveguide considered in §86

$$B_z = \left(\frac{n\pi K}{b}\right) \sin\frac{m\pi x}{a} \cos\frac{n\pi y}{b} \exp\{i(pt - kz)\},$$

$$B_y = -\left(\frac{m\pi K}{a}\right) \cos\frac{m\pi x}{a} \sin\frac{n\pi y}{b} \exp\{i(pt - kz)\},$$

$$B_z = 0,$$

and

$$E_x = -\left(\frac{km\pi c^2 K}{pa}\right) \cos\frac{m\pi x}{a} \sin\frac{n\pi y}{b} \exp\{i(pt-kz)\},$$

$$E_y = -\left(\frac{kn\pi c^2 K}{pb}\right) \sin\frac{m\pi x}{a} \cos\frac{n\pi y}{b} \exp\{i(pt-kz)\},$$

$$E_z = -\left(\frac{i\pi^2 c^2 K}{p}\right)\left(\frac{m^2}{a^2}+\frac{n^2}{b^2}\right) \sin\frac{m\pi x}{a} \sin\frac{n\pi y}{b} \exp\{i(pt-kz)\}.$$

17. Consider propagation in a cylindrical waveguide of radius a and take the z axis to coincide with the axis of the guide. By using cylindrical polar coordinates (r, θ, z) show that ϕ must be a solution of

$$\frac{1}{r}\frac{\partial}{\partial r}\left(r\frac{\partial\phi}{\partial r}\right) + \frac{1}{r^2}\frac{\partial^2\phi}{\partial\theta^2} + \varkappa^2\phi = 0.$$

8

General considerations

§88 Doppler effect

The speed at which waves travel in a medium is usually independent of the velocity of the source; thus, if a pebble is thrown into a pond with a horizontal velocity, the resulting water waves travel radially outwards from the centre of disturbance in the form of concentric circles, with a speed which is independent of the velocity of the pebble that caused them.

When we have a moving source, sending out waves continuously as it moves, the velocity of the waves is often unchanged, but the wavelength and frequency, as noted by a stationary observer, may be altered. The velocity is changed, however, when there is dispersion.

Thus, consider a source of waves moving towards an observer with velocity u. Then, since the source is moving, the waves which are between the source and the observer will be crowded into a smaller distance than if the source had been at rest. This is shown in Fig. 30, where the waves are drawn both for (a) a stationary and (b) a moving source. If the frequency is n, then in time t the source emits nt waves.

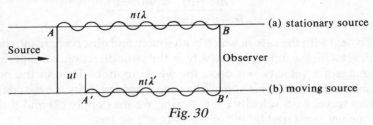

Fig. 30

If the source had been at rest, these waves would have occupied a length AB. But due to its motion the source has covered a distance ut, and hence these nt waves are compressed into a length $A'B'$, where $AB - A'B' = ut$. Thus

$$nt\lambda - nt\lambda' = ut,$$

and so

$$\lambda' = \lambda - u/n = \lambda(1 - u/c), \tag{1}$$

if c is the wave velocity. If the corresponding frequencies measured by the fixed observer are n and n', then, since $n\lambda = c = n'\lambda'$, we have

$$n' = \frac{nc}{c - u}. \tag{2}$$

If the source is moving towards the observer the frequency is increased; if it moves away from him, the frequency is decreased. This explains the sudden change of pitch noticed by a stationary observer when a train sounding its klaxon passes him. The actual change in this case is from $nc/(c - u)$ to $nc/(c + u)$, so that

$$\Delta n = 2ncu/(c^2 - u^2). \tag{3}$$

This phenomenon of the change of frequency when a source is moving is known as the **Doppler effect**. It applies equally well if the observer is moving instead of the source, or if both are moving.

For, consider the case of the observer moving with velocity v away from the source, which is supposed to be at rest. Let us superimpose upon the whole motion, observer, source and waves, a velocity $-v$. We shall then have a situation in which the observer is at rest, the source has a velocity $-v$, and the waves travel with a speed $c - v$. We may apply equation (2) which will then give the appropriate frequency as registered by the observer; if this is n'', then

$$n'' = \frac{n(c - v)}{(c - v) - (-v)} = \frac{n(c - v)}{c}. \tag{4}$$

To deal with the case in which both source and observer are moving, with velocities u and v, respectively, in the same direction, we superimpose again a velocity $-v$ upon the whole motion. Then in the new problem, the observer is at rest, the source has a velocity $u - v$, and the waves travel with velocity $c - v$. Again, we may apply (2) and if the frequency registered by the observer is n''', we have

$$n''' = \frac{n(c - v)}{(c - v) - (u - v)} = \frac{n(c - v)}{c - u}. \tag{5}$$

These considerations are of importance in acoustic and optical problems; it is not difficult to extend them to deal with cases in which the various velocities are not in the same line, but we shall not discuss such problems here.

§89   Beats

We have shown in Chapter 1, §6, that we may superpose any number of separate solutions of the wave equation. Suppose that we have two harmonic solutions (Chapter 1, equation (11)) with equal amplitudes and nearly equal frequencies. Then the total disturbance is

$$\phi = a \cos 2\pi(k_1 x - n_1 t) + a \cos 2\pi(k_2 x - n_2 t)$$

$$= 2a \cos 2\pi\left\{\frac{k_1 + k_2}{2}x - \frac{n_1 + n_2}{2}t\right\}\cos 2\pi\left\{\frac{k_1 - k_2}{2}x - \frac{n_1 - n_2}{2}t\right\}. \tag{6}$$

The first cosine factor represents a wave very similar to the original waves, whose frequency and wavelength are an average of the two initial values, and which moves with a velocity $(n_1 + n_2)/(k_1 + k_2)$. This is practically the same as the velocity of the original waves, and is indeed exactly the same if $n_1/k_1 = n_2/k_2$. But the second cosine factor, which changes much more slowly both with respect to x and t, may be regarded as a varying amplitude. Thus, for the resultant of the two original waves, we have a wave of approximately the same wavelength and frequency, but with an amplitude that changes both with time and distance.

We may represent this graphically, as in Fig. 31. The outer solid profile is the curve

$$y = 2a \cos 2\pi\left\{\frac{k_1 - k_2}{2}x - \frac{n_1 - n_2}{2}t\right\}.$$

The other profile curve is the reflection of this in the x axis. The actual disturbance ϕ lies between these two boundaries, cutting the axis of x at regular intervals, and touching alternately the upper and lower profile curves. If the velocities of the two component waves are the same, so that $n_1/k_1 = n_2/k_2$, then the wave system shown in Fig. 31

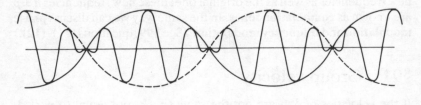

Fig. 31

moves steadily forward without change of shape. The case when n_1/k_1 is not equal to n_2/k_2 is dealt with in **§91**.

Suppose that ϕ refers to sound waves. Then we shall hear a resultant wave whose frequency is the mean of the two original frequencies, but whose intensity fluctuates with a frequency twice that of the solid profile curve. This fluctuating intensity is known as **beats**; its frequency, which is known as the **beat frequency**, is just $n_1 - n_2$, that is, the difference of the component frequencies. We can detect beats very easily with a piano slightly out of tune, or with two equal tuning-forks on the prongs of one of which we have put a little sealing wax to decrease its frequency. Determination of the beat frequency between a standard tuning-fork and an unknown frequency is one of the best methods of determining the unknown frequency. Low frequency beats are unpleasant to the ear.

§90   Amplitude modulation

There is another phenomenon closely related to beats. Let us suppose that we have a harmonic wave $\phi = A \cos 2\pi(nt - kx)$, with amplitude A and frequency n. Suppose further that the amplitude A is made to vary with the time in such a way that at $x = 0$, $A = a + b \cos 2\pi pt$. If the wave is to move with velocity $c = n/k$, ϕ must be some function of $ct - x$. So for the general x, $A = a + b \cos 2\pi p(t - kx/n)$. This is known as **amplitude modulation**. The result is

$$\phi = \{a + b \cos 2\pi p(t - kx/n)\} \cos 2\pi(nt - kx)$$

$$= a \cos 2\pi(nt - kx)$$

$$+ \frac{b}{2}\left\{\cos\left[2\pi(n+p)\left(t - \frac{kx}{n}\right)\right] + \cos\left[2\pi(n-p)\left(t - \frac{kx}{n}\right)\right]\right\}.$$

The effect of modulating, or varying, the amplitude, is to introduce two new frequencies as well as the original one; these new frequencies $n \pm p$ are known as **combination tones**. In the same way we can discuss **phase modulation** and **frequency modulation**. (See **§99**, questions (11)–(12).)

§91   Group velocity

If the velocities of **§89** are not the same (n_1/k_1 not equal to n_2/k_2), then the profile curves in Fig. 31 move with a speed $(n_1 - n_2)/(k_1 - k_2)$, which is different from that of the more rapidly oscillating part, whose speed is $(n_1 + n_2)/(k_1 + k_2)$. In other words, the individual waves in Fig. 31 advance through the profile, gradually increasing and then decreasing their amplitude, as they give place to other succeeding waves. This explains why, on the seashore, a wave which looks very large when it is some distance away from the shore, gradually reduces in height as it moves in, and may even disappear before it is sufficiently close to break.

This situation arises whenever the velocity of the waves, that is, their **wave velocity** V, is not constant, but depends on the frequency. We have already met this phenomenon which is known as **dispersion**. We deduce that in a dispersive system the only periodic wave profile that can be transmitted without change of shape is a single harmonic wave train; any other wave profile, which may be analysed into two or more harmonic wave trains, will change as it is propagated. The actual velocity of the profile curves in Fig. 31 is known as the **group velocity** U. We see from (6) that if the two components are not very different, $V = n/k$, and

$$U = (n_1 - n_2)/(k_1 - k_2) = \mathrm{d}n/\mathrm{d}k. \tag{7}$$

In terms of the wavelength λ, we have $k = 1/\lambda$, so that

$$U = \frac{\mathrm{d}n}{\mathrm{d}(1/\lambda)} = -\lambda^2 \frac{\mathrm{d}n}{\mathrm{d}\lambda}. \tag{8}$$

We could equally well write this

$$U = \frac{\mathrm{d}n}{\mathrm{d}k} = \frac{\mathrm{d}(kV)}{\mathrm{d}k} = V + k\frac{\mathrm{d}V}{\mathrm{d}k} = V - \lambda\frac{\mathrm{d}V}{\mathrm{d}\lambda}. \tag{9}$$

Our calculation has considered just two waves. But the form of equation (7) shows that we could equally well consider any number of waves superposed, provided that for any two of them $n_1 - n_2$ and $k_1 - k_2$ were sufficiently small for us to take their ratio constant and equal to $\mathrm{d}n/\mathrm{d}k$. If this condition is not satisfied we have to go to a closer approximation, as in §92.

The importance of group velocity lies in the fact that the energy is propagated with this velocity. We have already met several cases in which the wave velocity depends on the frequency; we shall calculate the group velocity for three of them.

Surface waves on a liquid of depth h. The analysis of Chapter 5, equation (32) shows that the velocity of surface waves on a liquid of depth h is given by

$$V^2 = \frac{g\lambda}{2\pi} \tanh \frac{2\pi h}{\lambda}.$$

According to (9) therefore, the group velocity is $V - \lambda \, dV/d\lambda$, so that

$$U = \tfrac{1}{2}V\left\{1 + \frac{4\pi h}{\lambda} \operatorname{cosech} \frac{4\pi h}{\lambda}\right\}. \tag{10}$$

When h is small, the two velocities are almost the same, but when h is large, $U = V/2$, so that the group velocity for deep sea waves is one-half of the wave velocity. Equation (10) is the same as the expression obtained §59, equation (47), for the rate of transmission of energy in these surface waves. Thus the energy is transmitted with the group velocity.

Electric waves in a dielectric medium.   The analysis in Chapter 7, §76, shows that the wave velocity in a dielectric medium is given by

$$V^2 = c^2 / \varepsilon_0 \mu_0 \varkappa_e \varkappa_m.$$

We may put $\varkappa_m = 1$ for waves in the visible region. Now the dielectric constant $\varkappa_e$ is not independent of the frequency, and so V depends on λ. The group velocity follows from (9); it is

$$U = V\left\{1 + \frac{\lambda}{2\varkappa_e} \frac{\partial \varkappa_e}{\partial \lambda}\right\}. \tag{11}$$

In most regions, especially when λ is long, $\varkappa_e$ decreases when λ increases so that U is less than V. For certain wavelengths, however, particularly those in the neighbourhood of a natural frequency of the atoms of the dielectric, there is anomalous dispersion, and U may exceed V. When λ is large, we have the approximate formula

$$\varkappa_e = A + B/\lambda^2 + C/\lambda^4.$$

It then appears from (11) that

$$U = V\frac{A - C/\lambda^4}{A + B/\lambda^2 + C/\lambda^4}.$$

Electric waves in a conducting medium.   The analysis in Chapter 7, §83, shows that the electric vector is propagated with an exponential term $\exp\{ip(t - \gamma z)\}$, where $\gamma^2 = \sigma\mu_0 \varkappa_m / 2p$. Thus $V^2 = 1/\gamma^2 = 2p/\sigma\mu_0 \varkappa_m$. According to (7), the group velocity is

$$U = \frac{dp}{d(p\gamma)} = \left(\gamma + p\frac{d\gamma}{dp}\right)^{-1}.$$

If we suppose that σ and $\varkappa_m$ remain constant for all frequencies, then this reduces to

$$U = 2/\gamma = 2V. \tag{12}$$

The group velocity here is actually greater than the wave velocity.

§92   Motion of wave packets

We shall now extend this discussion of group velocity to deal with the case of more than two component waves. We shall suppose that the wave profile is split up into an infinite number of harmonic waves of the type

$$\exp\{2\pi i(kx - nt)\}, \tag{13}$$

in which the wave number k has all possible values; we can suppose that the wave velocity depends on the frequency, so that n is a function of k. If the amplitude of the component wave (13) is $a(k)$ per unit range of k, then the full disturbance is

$$\phi(x, t) = \int_{-\infty}^{\infty} a(k)\exp\{2\pi i(kx - nt)\}\,dk. \tag{14}$$

This collection of superposed waves is known as a **wave packet**. The most interesting wave packets are those in which the amplitude is largest for a certain value of k, say k_0, and is vanishingly small if $k - k_0$ is large. Then the component waves mostly resemble $\exp\{2\pi i(k_0 x - n_0 t)\}$, and there are not many waves which differ greatly from this.

We shall discuss in detail the case in which

$$a(k) = A\exp\{-\sigma(k - k_0)^2\}. \tag{15}$$

This is known as a **Gaussian wave packet**, after the mathematician Gauss, who used the exponential function (15) in many of his investigations of other problems. A, σ and k_0 are, of course, constant for any one packet.

Let us first determine the shape of the wave profile at $t = 0$. The integral in (14) is much simplified because the term in n disappears. In fact,

$$\phi(x, 0) = \int_{-\infty}^{\infty} A \exp\{-\sigma(k - k_0)^2\} \cdot \exp(2\pi i k x)\, dk.$$

On account of the term $\exp\{-\sigma(k - k_0)^2\}$, the only range of k which contributes significantly to this integral lies around k_0; since when $k - k_0 = 1/\sqrt{\sigma}$ this term becomes e^{-1}, and for larger values of $k - k_0$ it becomes rapidly smaller, this range of k is of order of magnitude $\Delta k = 1/\sqrt{\sigma}$. In order to evaluate the integral, we use the result

$$\int_{-\infty}^{+\infty} \exp(au - bu^2)\, du = e^{a^2/4b} \int_{-\infty}^{+\infty} \exp\{-b(u^2 - au/b + a^2/4b^2)\}\, du$$

$$= e^{a^2/4b} \int_{-\infty}^{+\infty} \exp(-bv^2)\, dv$$

$$= \sqrt{(\pi/b)} \exp(a^2/4b). \tag{17}$$

This enables us to integrate at once, and we find that

$$\phi(x, 0) = A\sqrt{(\pi/\sigma)} \exp(-\pi^2 x^2/\sigma) \exp(2\pi i k_0 x). \tag{18}$$

The term $\exp(2\pi i k_0 x)$ represents a harmonic wave, whose wavelength $\lambda = 1/k_0$, and the other factors give a varying amplitude $A\sqrt{(\pi/\sigma)} \exp(-\pi^2 x^2/\sigma)$. If we take the real part of (18), $\phi(x, 0)$ has the general shape shown in Fig. 32. The outer curves in this figure are the two Gaussian curves

$$y = \pm A\sqrt{(\pi/\sigma)} \exp(-\pi^2 x^2/\sigma),$$

and $\phi(x, 0)$ oscillates between them. Our wave packet (14) represents, at $t = 0$, one large pulse containing several oscillations. If we define a **half-width** as the value of x that reduces the amplitude to $1/e$ times its maximum value, then the half-width of this pulse is $(\sqrt{\sigma})/\pi$.

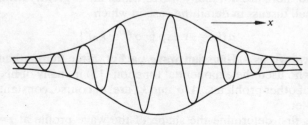

Fig. 32

At later times, $t > 0$, we have to integrate (14) as it stands. To do this we require a detailed knowledge of n as a function of k. If we expand according to Taylor's theorem, we can write

$$n = n_0 + \alpha(k - k_0) + \beta(k - k_0)^2/2 + \ldots,$$

where

$$\alpha = (dn/dk)_0, \qquad \beta = (d^2n/dk^2)_0 \ldots. \tag{19}$$

As a rule the first two terms are the most important, and if we neglect succeeding terms, we may integrate, using (17). The result is

$$\phi(x, t) = \int_{-\infty}^{+\infty} A \exp\{-\sigma(k - k_0)^2\}\exp\{2\pi i[kx - t(n_0 + a(k - k_0))]\}\, dk$$

$$= A\sqrt{(\pi/\sigma)} \exp\{-\pi^2(x - at)^2/\sigma\} . \exp\{2\pi i(k_0 x - n_0 t)\}. \tag{20}$$

When $t = 0$, it is seen that this does reduce to (18), thus providing a check upon our calculations. The last term in (20) shows that the individual waves move with a wave velocity n_0/k_0 but their boundary amplitude is given by the first part of the expression, namely, by $A\sqrt{(\pi/\sigma)} \exp\{-\pi^2(x - at)^2/\sigma\}$. Now this expression is exactly the same as in (18), drawn in Fig. 32, except that it is displaced a distance at to the right. We conclude, therefore, that the group *as a whole* moves with velocity $\alpha = (dn/dk)_0$, but that individual waves within the group have the wave velocity n_0/k_0. The velocity of the group as a whole is just what we have previously called the group velocity (7).

If we take one more term in (19) and integrate to obtain $\phi(x, t)$ we find that ϕ has the same form as in (20) except that σ is replaced by $\sigma + \pi\beta it$. The effect of this is twofold; in the first place it introduces a variable phase into the term $\exp\{2\pi i(k_0 x - n_0 t)\}$, and in the second place it changes the exponential term in the boundary amplitude curve to the form

$$\exp\left[\frac{-\pi^2\sigma(x - at)^2}{\sigma^2 + \pi^2\beta^2 t^2}\right].$$

This is still a Gaussian curve, but its half-width is increased to

$$\{(\sigma^2 + \pi^2\beta^2 t^2)/\sigma\pi^2\}^{1/2}. \tag{21}$$

We notice therefore that the wave packet moves with the wave velocity n_0/k_0, and group velocity $(dn/dk)_0$, spreading out as it goes in such a way that its half-width at time t is given by (21).

The importance of the group velocity lies mainly in the fact that in most problems where dispersion occurs, the group velocity is the velocity with which the energy is propagated. We have already met this in previous paragraphs.

§93   Kirchhoff's solution of the wave equation

We shall next give a general discussion of the standard wave equation

$$\nabla^2 \phi = \frac{1}{c^2} \frac{\partial^2 \phi}{\partial t^2},$$

in which c is constant. We shall show that the value of ϕ at any point P (which may, without loss of generality be taken to be the origin) may be obtained from a knowledge of the values of ϕ, $\dfrac{\partial \phi}{\partial n}$ and $\dfrac{\partial \phi}{\partial t}$ on any given closed surface S, which may or may not surround P; the values of ϕ and its derivatives on S have to be associated with times which differ somewhat from the time at which we wish to determine ϕ_P.

Let us analyse ϕ into components with different frequencies; each component itself must satisfy the wave equation, and by the principle of superposition, which holds when c is constant, we can add the various components together to obtain the full solution. Let us consider first that part of ϕ which is of frequency p: we may write it in the form

$$\psi(x, y, z) \exp (ikct), \tag{22}$$

where

$$k = 2\pi p/c. \tag{23}$$

ψ is the space part of the disturbance, and it satisfies the Helmholtz equation

$$(\nabla^2 + k^2)\psi = 0. \tag{24}$$

This last equation may be solved by using Green's theorem. This theorem states that if ψ_1 and ψ_2 are any two functions, and S is any closed surface, which may consist of two or more parts, such that ψ_1

and ψ_2 have no singularities inside it, then

$$\iint \{\psi_2 \nabla^2 \psi_1 - \psi_1 \nabla^2 \psi_2\}\, d\tau = \int \left\{\psi_2 \frac{\partial \psi_1}{\partial n} - \psi_1 \frac{\partial \psi_2}{\partial n}\right\} dS. \qquad (25)$$

The volume integral on the left-hand side is taken over the whole volume bounded by S, and $\partial/\partial n$ denotes differentiation along the outward normal to dS.

In this equation ψ_1 and ψ_2 are arbitrary, so we may put ψ_1 equal to ψ, the solution of (24), and $\psi_2 = (1/r) \exp(-ikr)$, r being measured radially from the origin P. We take the volume through which we integrate to be the whole volume contained between the given closed surface S shown in Fig. 33 and a small sphere Σ around the origin. We have to exclude the origin because ψ_2 becomes infinite at that point. Fig. 33 is drawn for the case of P within S; the analysis holds just as well if P lies outside S.

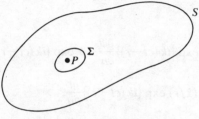

Fig. 33

Now it can easily be verified that $\nabla^2 \psi_2 = -k^2 \psi_2$, so that the left-hand side of (25) becomes

$$\iint \psi_2 (\nabla^2 + k^2) \psi\, d\tau,$$

and this vanishes, since $(\nabla^2 + k^2)\psi = 0$ by (24). The right-hand side of (25) consists of two parts, representing integrations over S and Σ. On Σ the outward normal is directed towards P and hence this part of the full expression is

$$\int \left\{(1/r) \exp(-ikr)\left(-\frac{\partial \psi}{\partial r}\right) - \psi\left(-\frac{\partial}{\partial r}[(1/r) \exp(-ikr)]\right)\right\} d\Sigma.$$

When we make the radius of Σ tend to 0, only one term remains; it is

$$-\int \psi(1/r^2) \exp(-ikr)\, d\Sigma = -\int \psi(1/r^2) \exp(-ikr)\, . \, r^2\, d\omega,$$

where $d\omega$ is an element of solid angle round P. Taking the limit as r tends to zero, this gives us a contribution $-4\pi\psi_P$. Equation (25) may therefore be written

$$4\pi\psi_P = \int \left\{ (1/r) \exp(-ikr) \frac{\partial \psi}{\partial n} - \psi \frac{\partial}{\partial n} [(1/r) \exp(-ikr)] \right\} dS$$

$$= \int \left\{ (1/r) \exp(-ikr) \frac{\partial \psi}{\partial n} - \psi \exp(-ikr) \frac{\partial}{\partial n} \left(\frac{1}{r} \right) \right.$$

$$\left. + ik\psi(1/r) \exp(-ikr) \frac{\partial r}{\partial n} \right\} dS.$$

Since by definition $\phi = \psi(x, y, z) \exp(ikct)$, we can write this last equation in the form

$$\phi_P = \frac{1}{4\pi} \int X \, dS, \qquad (26)$$

where

$$X = (1/r) \exp \{ik(ct - r)\} \frac{\partial \psi}{\partial n} - \psi \exp \{ik(ct - r)\} \frac{\partial}{\partial n} \left(\frac{1}{r} \right)$$

$$+ ik\psi(1/r) \exp \{ik(ct - r)\} \frac{\partial r}{\partial n},$$

$$= A - B + C, \text{ say.}$$

We may rewrite X in a simpler form; for on account of the time variation of ϕ, $\psi \exp \{ik(ct - r)\}$ is the same as ϕ taken not at time t, but at time $t - r/c$. If we write this symbolically $[\phi]_{t-r/c}$, then

$$B = \frac{\partial}{\partial n} \left(\frac{1}{r} \right) [\phi]_{t-r/c}.$$

In a similar way,

$$A = \frac{1}{r} \left[\frac{\partial \phi}{\partial n} \right]_{t-r/c},$$

and

$$C = \frac{1}{cr} \frac{\partial r}{\partial n} \left[\frac{\partial \phi}{\partial t} \right]_{t-r/c},$$

where, for example, $[\partial\phi/\partial n]_{t-r/c}$ means that we evaluate $\partial\phi/\partial n$ as a function of x, y, z, t and then replace t by $t - r/c$. We call $t - r/c$ the

retarded time. This retardation arises, of course, because of the finite speed of propagation of the wave and the dependence of the solution on the corresponding domain of dependence (cf. Chapter 1, §11). We have therefore proved that

$$\phi_P = \frac{1}{4\pi} \int X \, dS,$$

where

$$X = \frac{1}{r}\left[\frac{\partial \phi}{\partial n}\right]_{t-r/c} - \frac{\partial}{\partial n}\left(\frac{1}{r}\right)[\phi]_{t-r/c} + \frac{1}{cr}\frac{\partial r}{\partial n}\left[\frac{\partial \phi}{\partial t}\right]_{t-r/c} \qquad (27)$$

So far we have been dealing with waves of one definite frequency. But there is nothing in (27) which depends upon the frequency, and hence, by summation over all the components for each frequency present in our complete wave, we obtain a result exactly the same as (27) but without the restriction to a single frequency.

This theorem, which is due to Kirchhoff, is of great theoretical importance; for it implies (a) that the value of ϕ may be regarded as the sum of contributions $X/4\pi$ from each element of area of S; this may be called the law of addition of small elements, and is familiar in a slightly different form in optics as **Huygens' Principle**; and (b) that the contribution of dS depends on the value of ϕ, not at time t, but at time $t - r/c$. Now r/c is the time that a signal would take to get from dS to the point P, so that the contribution made by dS depends not on the present value of ϕ at dS, but on its value at that particular previous moment when it was necessary for a signal to leave dS in order that it should just have arrived at P. This is the justification for the title of retarded time, and for this reason also, $[\phi]_{t-r/c}$ is sometimes known as a **retarded potential**.

It is not difficult to verify that we could have obtained a solution exactly similar to the above, but involving $t + r/c$ instead of $t - r/c$; we should have taken ψ_2 in the previous work to be $(1/r)\exp{(ikr)}$ instead of $(1/r)\exp{(-ikr)}$. In this way we should have obtained **advanced potentials**, $[\phi]_{t+r/c}$, and advanced times, instead of retarded potentials and retarded times. More generally, too, we could have superposed the two types of solution, but we shall not discuss this matter further.

In the case in which $c = \infty$, so that signals have an infinite velocity, the fundamental equation reduces to Laplace's equation, $\nabla^2 \phi = 0$, and the question of time variation does not arise. Our equation (27) reduces to the standard solution for problems of electrostatics.

§94 Fresnel's principle

We shall apply this theory to the case of a source O sending out spherical harmonic waves, and we shall take S to be a closed surface surrounding the point P at which we want to calculate ϕ, as shown in Fig. 34. Consider a small element of dS at Q; the outward normal

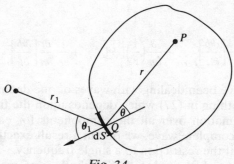

Fig. 34

makes angles θ_1 and θ with QO and PQ, and these two distances are r_1 and r, respectively. The value of ϕ and Q is given by the form appropriate to a spherical wave (see Chapter 1, equation (24)):

$$\phi_Q = \frac{a}{r_1} \cos m(ct - r_1). \tag{28}$$

Thus

$$\frac{\partial \phi}{\partial n} = -\cos \theta_1 \frac{\partial \phi}{\partial r_1}$$

$$= a \cos \theta_1 \left\{ \frac{1}{r_1^2} \cos m(ct - r_1) - \frac{m}{r_1} \sin m(ct - r_1) \right\}.$$

Now $\lambda = 2\pi/m$, so that if r_1 is much greater than λ, which will almost always happen in practical problems, we may put

$$\frac{\partial \phi}{\partial n} = -\frac{ma \cos \theta_1}{r_1} \sin m(ct - r_1).$$

Also

$$\frac{\partial}{\partial n}\left(\frac{1}{r}\right) = -\frac{1}{r^2} \cos \theta$$

and

$$\frac{\partial \phi}{\partial t} = -\frac{amc}{r_1} \sin m(ct - r_1).$$

The retarded values are easily found, and in fact, from (27),

$$X = -\frac{ma \cos \theta_1}{rr_1} \sin m(ct - [r + r_1])$$

$$+ \frac{a}{r^2 r_1} \cos \theta \cos m(ct - [r + r_1])$$

$$- \frac{1}{cr} \frac{amc}{r_1} \cos \theta \sin m(ct - [r + r_1]).$$

We may neglect the second terms on the right if r_1 is much greater than λ, and so

$$X = -\frac{ma}{rr_1}(\cos \theta + \cos \theta_1) \sin m(ct - [r + r_1]). \qquad (29)$$

Combining (29) with (26) it follows that

$$\phi_P = -\frac{1}{4\pi} \int \frac{ma}{rr_1}(\cos \theta + \cos \theta_1) \sin m(ct - [r + r_1]) \, dS$$

$$= -\frac{a}{2\lambda} \int \frac{1}{rr_1}(\cos \theta + \cos \theta_1) \sin m(ct - [r + r_1]) \, dS. \qquad (30)$$

If, instead of a spherical wave, we had had a plane wave coming from the direction of O, we should write

$$\phi_P = -a \cos m(ct - r_1),$$

r_1 now being measured from some plane perpendicular to OQ, and (30) would be changed to

$$\phi_P = -\frac{a}{2\lambda} \int \frac{1}{r}(\cos \theta + \cos \theta_1) \sin m(ct - [r + r_1]) \, dS. \qquad (31)$$

We may interpret (30) and (31) as follows. The effect at P is the same as if each element dS sends out a wave of amplitude

$$\frac{A}{\lambda r}\left(\frac{\cos \theta + \cos \theta_1}{2}\right) dS,$$

A being the amplitude of the incident wave at dS; further, these waves are a quarter of a period in advance of the incident wave, as is shown by the term $-\sin m(ct - [r + r_1])$ instead of $\cos m(ct - r_1)$. $\frac{1}{2}(\cos\theta + \cos\theta_1)$ is called the **inclination factor** and if, as often happens, only small values of θ and θ_1 occur significantly, it has the value unity. This interpretation of (30) and (31) is known as **Fresnel's principle**.

The presence of this inclination factor removes a difficulty which was inherent in Huygens' principle; this principle is usually stated in the form that each element of a wave-front emits wavelets in all directions, and these combine to form the observed progressive wave-front. In such a statement there is nothing to show why the wave does not progress backwards as well as forwards, since the wavelets should combine equally in either direction. The explanation is, of course, that for points behind the wavefront $\cos\theta$ is negative with a value either exactly or approximately equal to $-\cos\theta_1$, and so the inclination factor is small. Each wavelet is therefore propagated almost entirely in the forward direction.

Now let us suppose that some screens are introduced, and that they cover part of the surface of S. If we assume that the distribution of ϕ at any point Q near the screens is the same as it would have been if the screens were not present, we have merely to integrate (30) or (31) over those parts of S which are not covered. This approximation, which is known as **St. Venant's principle**, is not rigorously correct, for there will be distortions in the value of ϕ_Q extending over several wavelengths from the edges of each screen. It is, however, an excellent approximation for most optical problems, where λ is small; indeed (30) and (31) form the basis of the whole theory of the diffraction of light. With sound waves, on the other hand, in which λ is often of the same order of magnitude as the size of the screen, it is only roughly correct.

§95   Diffraction at a pin hole

Let us illustrate this discussion with an example of the analysis summarised in (31). Consider an infinite screen shown in Fig. 35 which we may take to be the (x, y) plane. A small part of this screen (large compared with the wavelength of the waves but small compared with other distances involved) is cut away, leaving a hole through which waves may pass. We suppose that a set of plane harmonic waves is travelling in the positive z direction, and falls on the screen; we want to find the resulting disturbance at a point P behind the screen.

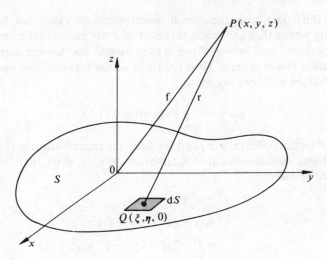

Fig. 35

In accordance with §94 we take the surface of S to be the infinite (x, y) plane, completed by the infinite hemisphere on the positive side of the (x, y) plane. We may divide the contributions to (31) into three parts. The first part arises from the aperture, the second part arises from the rest of the screen, and the third part arises from the hemisphere.

If the incident harmonic waves are represented by $\phi = a \cos m(ct - z)$ this first contribution amounts to

$$-\frac{a}{2\lambda} \int \frac{1}{r} (1 + \cos \theta) \sin m(ct - r) \, dS.$$

We have put $\theta_1 = 0$ in this expression since the waves fall normally onto the (x, y) plane. We shall only be concerned here with points P which lie behind, or nearly behind, the aperture so that we may also put $\cos \theta = 1$ without loss of accuracy. This contribution is then

$$-\frac{a}{\lambda} \int \frac{1}{r} \sin m(ct - r) \, dS. \tag{32}$$

The second part, which comes from the remainder of the (x, y) plane vanishes, since no waves penetrate the screen and thus there are no secondary waves starting there.

The third part, from the infinite hemisphere, also vanishes, because the only waves that can reach this part of S are those that came from the aperture, and when these waves reach the hemisphere their inclination factor is zero. Thus (32) is in actual fact the only non-zero contribution and we may write

$$\phi_P = -\frac{a}{\lambda} \int \frac{1}{r} \sin m(ct-r) \, \mathrm{d}S. \tag{33}$$

Let P be the point (x, y, z) and consider the contribution to (33), that arises from a small element of the aperture at $Q(\xi, \eta, 0)$. If $OP = f$, and $QP = r$, we have

$$f^2 = x^2 + y^2 + z^2,$$
$$r^2 = (x-\xi)^2 + (y-\eta)^2 + z^2$$
$$= f^2 - 2x\xi - 2y\eta + \xi^2 + \eta^2. \tag{34}$$

Let us make the assumption that the aperture is so small that ξ^2/f^2 and η^2/f^2 may be neglected. Then to this approximation (34) shows us that

$$r = f - \frac{x\xi + y\eta}{f}.$$

So

$$\phi_P = -\frac{a}{\lambda} \int \frac{1}{r} \sin m\left(ct - f + \frac{x\xi + y\eta}{f}\right) \mathrm{d}S.$$

Again without loss of accuracy, to the approximation to which we are working, we may put $1/r = 1/f$, and then we obtain

$$\phi_P = -A \sin\{m(ct-f) + \varepsilon\},$$

where

$$A^2 = C^2 + S^2, \qquad \tan \varepsilon = S/C,$$

and

$$C(x, y) = \frac{a}{\lambda f} \int \cos \frac{2\pi}{\lambda f}(x\xi + y\eta) \, \mathrm{d}\xi \, \mathrm{d}\eta,$$

$$S(x, y) = \frac{a}{\lambda f} \int \sin \frac{2\pi}{\lambda f}(x\xi + y\eta) \, \mathrm{d}\xi \, \mathrm{d}\eta. \tag{35}$$

Once we know the shape of the aperture it is an easy matter to evaluate these integrals. Thus, if we consider the case of a rectangular aperture bounded by the lines $\xi = \pm\alpha$, $\eta = \pm\beta$, we soon verify that $S = 0$, and that

$$C = \frac{a}{\lambda f} \int_{-\alpha}^{+\alpha} \int_{-\beta}^{+\beta} \cos \frac{2\pi}{\lambda f}(x\xi + y\eta)\, \mathrm{d}\eta\, \mathrm{d}\xi$$

$$= \frac{4a}{\lambda f} \frac{\sin p\alpha x}{px} \frac{\sin p\beta y}{py}, \tag{36}$$

where $p = 2\pi/\lambda f$. If we are dealing with light waves, then the intensity is proportional to C^2 and the diffraction pattern thus observed in the plane $z = f$ consists of a grill network, with zero intensity corresponding to the values of x and y satisfying either $\sin p\alpha x = 0$, or $\sin p\beta y = 0$ but excluding $x = 0$ and $y = 0$.

§96 Fraunhofer diffraction theory

The discussion of the last paragraph related to the case of plane waves falling normally on an aperture whose size, while large compared with the wavelength, was still small compared with the distance from the aperture to the screen on which the pattern was being observed. We might refer to this as *diffraction at a pin-hole*. But the equations (35) arise in another far more important way which we must now explain, and which is known as **Fraunhofer Diffraction**.

Consider a plane wave shown as AA' in Fig. 36 falling normally on a convergent lens L. (L now replaces the previous pin-hole). This lens will convert the plane wave into a spherical wave which converges at Z, the focus. On account of the finite size of the lens the focus is not perfect, and we ask the question: what will be the intensity observed at a point P in the focal plane through Z?

To answer this question it is convenient to draw the wavefront ROR' of a wave that has just left the lens. We may regard this as part of a spherical surface with centre Z and radius equal to the focal length f. If we take O as origin and OZ as axis of Z, then the coordinates (ξ, η, ζ) of any point Q on the surface satisfy the equation

$$\xi^2 + \eta^2 + (\zeta - f)^2 = f^2,$$

or

$$\xi^2 + \eta^2 + \zeta^2 = 2f\zeta. \tag{37}$$

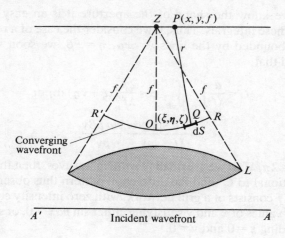

Fig. 36

Now by reasoning very similar to that used in §95 we may argue that if P is near Z, the total effect at P is just the sum of separate contributions arising from all the elements dS within the curved wavefront RR'. Let us suppose that the inclination factor may be put equal to unity, and that the amplitude at all points on RR' is a; this is the same as the amplitude in the incident wave AA'. Thus

$$\phi_Q = a \cos m(ct + f). \tag{38}$$

Let us write (x, y, f) for the coordinates of the point P at which the observation is made, and put $QP = r$. Then the appropriate form of (30) is

$$\phi_P = -\frac{a}{\lambda} \int \frac{1}{r} \sin m(ct - r + f)\, dS. \tag{39}$$

We may replace $1/r$ by $1/f$ in the first part of this integral: and note that

$$r^2 = (x - \xi)^2 + (y - \eta)^2 + (f - \zeta)^2$$
$$= f^2 + x^2 + y^2 - 2x\xi - 2y\eta \quad \text{by (37).}$$

In cases where this type of diffraction is important, f is large and x and y are small. We may therefore write

$$r^2 = f^2 - 2x\xi - 2y\eta,$$

so that effectively

$$r = f - \frac{x\xi + y\eta}{f}. \tag{40}$$

Combining (39) and (40) we have

$$\phi_P = -\frac{a}{\lambda f} \int \sin m\left(ct + \frac{x\xi + y\eta}{f}\right) dS.$$

Thus, on integrating,

$$\phi_P = -A \sin (mct + \varepsilon), \tag{41}$$

where A and ε are given by precisely the same formulae as in (35).

This kind of analysis will apply particularly to the image of a star in a telescope of long focal length. The star is so far away that it may be regarded as giving out a beam of parallel light. We have just shown therefore that the image of the star is not a point, but a pattern with maxima and minima, depending on the shape and size of the lens. For example, if a rectangular aperture (bounded by $\xi = \pm\alpha$, $\eta = \pm\beta$) is placed immediately behind the lens L, the diffraction pattern is a grill network, as in (36), and a circular aperture (Example 9 at the end of the chapter) gives diffraction rings around Z. In any case the finite extent of the central maximum, or zone, will put limits to our power of resolving the light from two close stars. For if the geometrical images of the two stars lie within one another's central zones, we shall experience difficulty in distinguishing whether there is really only one star, or two. But there is no space here to deal with this important matter any more closely.

§97   Retarded potential theory

We now offer a brief discussion of the equation

$$\nabla^2 \phi = \frac{1}{c^2} \frac{\partial^2 \phi}{\partial t^2} - 4\pi\rho, \tag{42}$$

where ρ is some given function of x, y, z and t. When $\rho = 0$ this is the standard wave equation, whose solution was discussed in §93. Equation (42) has already occurred in the propagation of electric waves when charges were present (Chapter 8, equations (17′) and (18′)). We may solve this equation in a manner very similar to that used in §93.

Thus, suppose that $\rho(x, y, z, t)$ is expressed in the form of a Fourier series with respect to t, namely,

$$\rho(x, y, z, t) = \sum a_k(x, y, z) \exp(ikct). \qquad (43)$$

There may be a finite, or an infinite number, of different values of k, and instead of a summation over discrete values of k we could, if we desired, include also an integration over a continuous range of values. We shall discuss here the case of discrete values of k; the reader will easily adapt our method of solution to deal with a continuum.

Suppose that $\phi(x, y, z, t)$ is itself analysed into components similar to (43), and let us write, similarly to (22),

$$\phi(x, y, z, t) = \sum_k \psi_k(x, y, z) \exp(ikct), \qquad (44)$$

the values of k being the same as in (43). If we substitute (43) and (44) into (42), and then equate coefficients of $\exp(ikct)$, we obtain an equation for ψ_k. It is

$$(\nabla^2 + k^2)\psi_k = -4\pi a_k. \qquad (45)$$

This equation may be solved just as in §93. Using Green's theorem as in (25), we put $\psi_1 = \psi_k(x, y, z)$, $\psi_2 = (1/r) \exp(-ikr)$, taking Σ and S to be the same as in Fig. 33. With these values, it is easily seen that the left-hand side of (25) no longer vanishes, but has the value

$$-4\pi \int \frac{a_k(x, y, z)}{r} \exp(-ikr) \, d\tau, \qquad (46)$$

the integral being taken over the space between Σ and S. The right-hand side may be treated exactly as in §93, and gives two terms, one due to integration over Σ, and the other to integration over S. The first of these is

$$-4\pi\psi_k(x_P, y_P, z_P). \qquad (47)$$

The second may be calculated just as on p. 159. Gathering the various terms together, we obtain

$$\psi_k(x_P, y_P, z_P) = \int \frac{a_k(x, y, z)}{r} \exp(-ikr) \, d\tau$$

$$+ \frac{1}{4\pi} \int \left\{ (1/r) \exp(-ikr) \frac{\partial \psi_k}{\partial n} - \psi_k \exp(-ikr) \frac{\partial}{\partial n} \left(\frac{1}{r} \right) \right.$$

$$\left. + ik\psi_k(1/r) \exp(-ikr) \frac{\partial r}{\partial n} \right\} \, dS. \qquad (48)$$

Combining (43), (44) and (48) we can soon verify that our solution can be written in the form

$$\phi(x_P, y_P, z_P) = \int \frac{[\rho]_{t-r/c}}{r} \, d\tau + \frac{1}{4\pi} \int X \, dS, \tag{49}$$

where X is defined by (27). This solution reduces to (27) in the case where $\rho = 0$, while it reduces to the well-known solution of electrostatics in the case where $c = \infty$.

We have now obtained the required solution of (42). Often, however, there will be conditions imposed by the physical nature of our problem that allow us to simplify (49). Thus, if $\rho(x, y, z, t)$ is finite in extent, and has only had non-zero values for a finite time $t > t_0$, we can make $X = 0$ by taking S to be the sphere at infinity. This follows because X is measured at the retarded time $t - r/c$, and if r is large enough, we shall have $t - r/c < t_0$, so that $[\phi]_{t-r/c}$ and its derivatives will be identically zero on S. In such a case we have the simple result

$$\phi(x_P, y_P, z_P) = \int \frac{[\rho]_{t-r/c}}{r} \, d\tau, \tag{50}$$

the integration being taken over the whole of space. Retarded potentials calculated in this way are very important in the Classical theory of electrons.

§98   Wave propagation in an inhomogeneous medium

So far, all our analysis has been based on the assumption that the wave propagation speed c that occurs in the wave equation is constant. There are, however, many situations in which the equation provides a suitable description of some physical phenomenon provided the wave speed c is allowed to vary with position. This happens, for example, in optics when light travels through a medium with a variable refractive index, in acoustics when sound waves travel through any inhomogeneous continuum which may be gaseous, liquid or solid, and when electromagnetic waves travel through an ionized gas.

Because of the complexity of the subject, we can offer no more than a brief outline of an important approximate method of approach that may often be used to resolve these problems. It is called the WKBJ method after Wentzel, Kramers, Brillouin and Jeffreys who first developed it. For simplicity, we shall discuss it in the context of the one-dimensional wave equation

$$\frac{\partial^2 \phi}{\partial x^2} = \frac{1}{c^2(x)} \frac{\partial^2 \phi}{\partial t^2}. \tag{51}$$

We already know that when $c(x) = c$ is constant, a wave travelling to the right with speed c and frequency $p/2\pi$ will be a solution of (51) if

$$\phi = A \exp\{ip(x/c - t)\}. \tag{52}$$

Our approach will be to seek a generalisation of this form of solution that will be appropriate to (51) when $c(x)$ is a slowly varying function relative to some reference speed, say $c(x_0)$, where x_0 is a convenient reference point. The form of solution that will be adopted as the analogue of (52) will be

$$\phi(x, t) = A(x) \exp\{i(Z(x) - pt)\}, \tag{53}$$

where $A(x)$ and $Z(x)$ are slowly varying functions of x that are to be determined.

Substitution of (53) into (51) followed by the cancellation of the common exponential factor leads to the result

$$\frac{d^2 A}{dx^2} + \left[\frac{p^2}{c^2(x)} - \left(\frac{dZ}{dx}\right)^2\right] A + i\left(2\frac{dA}{dx}\frac{dZ}{dx} + A\frac{d^2 Z}{dx^2}\right) = 0. \tag{54}$$

This immediately implies the pair of simultaneous ordinary differential equations

$$\frac{d^2 A}{dx^2} + \left[\frac{p^2}{c^2(x)} - \left(\frac{dZ}{dx}\right)^2\right] A = 0, \tag{55}$$

and

$$2\frac{dA}{dx}\frac{dZ}{dx} + A\frac{d^2 Z}{dx^2} = 0. \tag{56}$$

In general, we cannot hope to solve these exactly, but an approximate solution may be obtained as follows. We begin with equation (56) which can be expressed in a more convenient form without any approximation. Setting $q = dZ/dx$ it becomes

$$2q\frac{dA}{dx} + A\frac{dq}{dx} = 0,$$

or

$$2 \int_{x_0}^{x} \frac{dA}{A} + \int_{q_0}^{q} \frac{ds}{s} = 0.$$

Integration then gives

$$2 \log A(x) - 2 \log A(x_0) + \log\left(\frac{dZ}{dx}\right) - \log\left(\frac{dZ}{dx}\right)_{x=x_0} = 0,$$

which in turn yields

$$A(x) = A(x_0)[Z'(x_0)/Z'(x)]^{1/2}, \tag{57}$$

with the prime signifying differentiation with respect to x.

Turning our attention now to equation (55), let us assume that $A(x)$ is a slowly varying function for which we may neglect d^2A/dx^2. It then follows that to a first approximation

$$\frac{dZ}{dx} = \frac{p}{c(x)}, \tag{58}$$

which is equivalent to

$$Z(x) = \int_{x_0}^{x} \frac{p}{c(s)} \, ds. \tag{59}$$

Employing this approximation for $Z(x)$ in (57) gives

$$A(x) = A(x_0)\left[\frac{c(x)}{c(x_0)}\right]^{1/2}. \tag{60}$$

Substitution of (59) and (60) into (53) then yields the approximate solution

$$\phi(x, t) = A(x_0)\left[\frac{c(x)}{c(x_0)}\right]^{1/2} \exp\left\{i\left(\int_{x_0}^{x} \frac{p}{c(s)} \, ds - pt\right)\right\}. \tag{61}$$

To estimate the range of validity of this expression it is necessary to make more precise the nature of the slow variation of $c(x)$ with x. However, the source of the error was that when deriving expression (58) we neglected d^2A/dx^2 in (55). So, to estimate the effect this has, let us now suppose that the true form of dZ/dx can be written

$$\frac{dZ}{dx} = \frac{p}{c(x)}(1 + h(x)), \tag{62}$$

where $h(x)$ is the correction needed to make this result exact.

Substituting (60) and (62) into (55) gives

$$\frac{1}{2}\frac{d^2c}{dx^2} - \frac{1}{4}c\left(\frac{dc}{dx}\right)^2 - \frac{p^2}{c}(2h + h^2) = 0.$$

This expression also requires simplification, and to achieve it we assume the slow variation of $c(x)$ to be such that $(dc/dx)^2 \gg d^2c/dx^2$. Consequently, neglecting this second derivative together with $h^2(x)$ which we have taken to be small relative to $h(x)$, gives

$$|h(x)| = \frac{1}{8p^2}\left(\frac{dc}{dx}\right)^2. \tag{63}$$

The approximation will thus be valid provided $|h(x)|$ as defined here is such that $|h(x)| \ll 1$.

A corresponding solution to (61) exists for a wave travelling to the left and the approximate general solution will be the sum of these two solutions.

Result (61) shows that the amplitude of this travelling wave varies like $\sqrt{c(x)}$, while the term x/c in (52) is replaced by

$$\int_{x_0}^{x} \frac{ds}{c(s)}.$$

§99   Examples

1. An observer who is at rest notices that the frequency of a car appears to drop from 272 to 256 per second as the car passes him. Show that the speed of the car is approximately 36·2 km per hour. How fast must he travel in the direction of the car for the apparent frequency to rise to 280 per second and what would it drop to in that case as the car passed him.

2. Show that in the Doppler effect, when the source and observer are not moving in the same direction, the formulae of §88 are valid to give the various changes in frequency, provided that u and v denote, not the actual velocities, but the components of the two velocities along the direction in which the waves reach the observer.

3. The amplitude A of a harmonic wave $A \cos 2\pi(nt - kx)$ is modulated so that $A = a + b \cos 2\pi pt + c \cos^2 2\pi pt$. Show that combination tones of frequencies $n \pm p$, $n \pm 2p$ appear, and calculate their partial amplitudes.

4. The dielectric constant of a certain gas varies with the wavelength according to the law $\varkappa_e = A + B/\lambda^2 - C\lambda^2$, where A, B and C are constants. Show that the group velocity U of electromagnetic waves is given in terms of the wave velocity V by the formula

$$U = V\frac{A - 2C\lambda^2}{A + (B/\lambda^2) - C\lambda^2}.$$

5. In a region of anomalous dispersion (§91), the dielectric constant obeys the approximate law

$$\varkappa_e = 1 + \frac{A\lambda^2}{\lambda^2 - \lambda_0^2}.$$

A more accurate expression is

$$\varkappa_e = 1 + \frac{A\lambda^2(\lambda^2 - \lambda_0^2)}{(\lambda^2 - \lambda_0^2)^2 + B\lambda^2},$$

where A, B and λ_0 are constants. Find the group velocity of electric waves in these two cases.

6. Calculate the group velocity for ripples on an infinitely deep lake. (§61), equation (54).)

7. Investigate the motion of a wavepacket (§92) for which the amplitude a is given in terms of the wave number k by the relation

$$a(k) = 1 \quad \text{if } |k - k_0| < k_1$$
$$= 0 \text{ otherwise,}$$

k_0 and k_1 being constants. Assume that only the first two terms of the Taylor expansion of n in terms of k are required. Show that at time t the disturbance is

$$\phi(x, t) = \frac{\sin\{2\pi(x - at)k_1\}}{\pi(x - at)}\exp\{2\pi i(k_0 x - n_0 t)\},$$

where $a = (dn/dk)_0$. Verify that the wavepacket moves as a whole with the velocity a.

8. Show that when dS is normal to the incident light (§94), the inclination factor is $(1 + \cos\theta)/2$. Plot this function against θ, and thus show that each element dS of a wave gives zero amplitude immediately behind the direction of wave motion. Using the fact that the energy is proportional to the square of the amplitude of ϕ, show that, taken

alone, each element sends out $\frac{7}{8}$ of its energy forwards in front of the wave, and only $\frac{1}{8}$ backwards.

9. A plane wave falls normally on a small circular aperture of radius b. Discuss the pattern observed at a large distance f behind the aperture. Show that with the formulae of §95, if the incident wave is $\phi = a \cos m(ct - z)$, then $S = 0$, and if P is the point $(x, 0, f)$, then

$$c = \frac{2a}{\lambda f} \int_{-b}^{b} \sqrt{(b^2 - \xi^2)} \cdot \cos p\xi \, \mathrm{d}\xi \quad \text{where } p = 2\pi x/\lambda f,$$

$$= \frac{4ab^2}{\lambda f} \int_0^{\pi/2} \cos(pb \cos \theta) \sin^2 \theta \, \mathrm{d}\theta.$$

Expand $\cos(pb \cos \theta)$ in a power series in $\cos \theta$, and hence show that

$$C = \frac{\pi ab^2}{\lambda f} \left\{ 1 - \frac{1}{2}\left(\frac{k}{1!}\right)^2 + \frac{1}{3}\left(\frac{k^2}{2!}\right)^2 - \frac{1}{4}\left(\frac{k^3}{3!}\right)^2 + \dots \right\},$$

where $k = pb/2 = \pi bx/\lambda f$. Since the system is symmetrical around the z axis, this gives the disturbance at any point in the plane $z = f$. It can be shown that the infinite series is in fact a Bessel function of order unity. It gives rise to diffraction rings of diminishing intensity for large values of x.

10. The total charge q on a conducting sphere of radius a is made to vary so that $q = 4\pi a^2 \sigma$, where $\sigma = 0$ for $t < 0$, and $\sigma = \sigma_0 \sin pt$ for $t > 0$. Show that if $\varkappa_e = \varkappa_m = 1$, (§98 equation (18′)) the electric potential ϕ at a distance R from the centre of the sphere is given by

$$\phi = 0 \quad \text{for } ct < R - a,$$

$$\phi = \frac{2\pi ac\sigma_0}{pR}\{1 - \cos p[t - (R - a)/c] \quad \text{for } R - a < ct < R + a,$$

$$\phi = \frac{4\pi ac\sigma_0}{pR} \sin \frac{pa}{c} \sin p(t - R/c) \quad \text{for } R + a < ct.$$

11. The wave represented by $\phi = A \cos 2\pi(nt - kx + \varepsilon)$ suffers phase modulation in which $\varepsilon = a + b \cos 2\pi pt$. a, b and p are constants, and b^2 may be neglected. Show that in addition to the wave of given frequency n and amplitude A, combination tones appear, with frequency $n \pm p$, and amplitude πAb.

12. The wave represented by $\phi = A \cos 2\pi(nt - kx + \varepsilon)$ suffers frequency modulation in which $n = n_0 + a \cos 2\pi pt$. a, n_0 and p are

constants, a^2 may be neglected and $p \ll n_0$. Show that in addition to a wave of frequency n_0 and amplitude A, there are four combination tones of frequency $n_0 \pm p \pm a$ and amplitude $A/4$. It may be assumed that at is small.

13. Consider the ordinary differential equation

$$\frac{d^2 X}{dx^2} + f(x)X = 0$$

in which $f(x)$ is a slowly varying function of x. Modify the WKBJ method to show that when a solution is sought in the form

$$X(x) = \exp(iZ(x)),$$

then neglecting $d^2 Z/dx^2$ leads to the first approximation

$$Z(x) = \pm \int f^{1/2}\, dx.$$

Hence, by using this result to approximate the neglected term $d^2 Z/dx^2$, show that the second approximation is

$$Z(x) = \pm \int f^{1/2}\, dx + \frac{i}{4} \log f,$$

and the corresponding approximate general solution is

$$X(x) = \left\{ A \exp\left[i \int f^{1/2}\, dx \right] + B \exp\left[-i \int f^{1/2}\, dx \right] \right\} \Big/ [f(x)]^{1/4}.$$

14. Waves propagate in the (x, z) plane in an inhomogeneous medium in which the wave propagation speed $c = c(x)$ is a function only of the penetration x into the medium. By assuming a solution of the type

$$\phi(x, z, t) = X(x) \exp\{i(\kappa z - pt)\},$$

show how the result of the previous question may be used to find the approximate variation of the amplitude factor $X(x)$.

Nonlinear waves

§100   Nonlinearity and quasilinear equations

The linearity of a partial differential equation implies that any linear combination of solutions of the equation will also be a solution. This fundamental fact was first commented upon in §6 when introducing superposibility of solutions and then, more generally, in §7 in justification of the method of seperation of variables. In the event that a partial differential equation is nonlinear this property is lost, and it becomes impossible to employ separation of variable techniques, or any other argument that depends on superposibility.

Nonlinearity may enter a differential equation in many different ways. For example, in our derivation of the wave equation for a string in §17, we first obtained the nonlinear second order equation

$$\frac{\partial^2 y}{\partial t^2} = c^2 \left\{ 1 + \left(\frac{\partial y}{\partial x} \right)^2 \right\}^{-2} \frac{\partial^2 y}{\partial x^2}. \tag{1}$$

Here the right-hand side is a nonlinear function of y and it only becomes linear when we may neglect the term $\left(\frac{\partial y}{\partial x} \right)^2$. The situation is somewhat different when studying the mechanics of compressible fluids, for in the simplest case there are two simultaneous first order equations that govern the flow in three dimensions:

conservation of mass

$$\frac{\partial \rho}{\partial t} + \text{div} \, (\rho \mathbf{u}) = 0, \tag{2}$$

equation of motion

$$\frac{\partial \mathbf{u}}{\partial t} + \mathbf{u} \cdot \text{grad} \, \mathbf{u} + \frac{1}{\rho} \, \text{grad} \, p = \mathbf{0}. \tag{3}$$

These relate the fluid velocity $\mathbf{u}$, the density ρ and the pressure p. When there is an adiabatic equation of state $p = f(\rho)$, with f a known function characterising the fluid in question. This last relationship between p and ρ has already been used in the specific form

$$p = k\rho^{\gamma} \tag{4}$$

in Chapter 6, §63, equation (1) when discussing sound waves in an adiabatic gas.

In this particular system of simultaneous partial differential equations nonlinearity enters in various forms. The term div $(\rho\mathbf{v})$ is nonlinear because of the product of the two dependent variables ρ and $\mathbf{u}$, while $\mathbf{u}$. grad $\mathbf{u}$ is nonlinear because $\mathbf{u}$ and its derivatives are multiplied together. Finally the term $(1/\rho)$ grad p is nonlinear both because of the factor $1/\rho$ and because of the relationship between p and ρ given in (4).

Naturally, the study of a system such as the one comprising equations (2) and (3) is more complicated than the study of the single equation (1), but although apparently very different, both have, in fact, more in common than is at first apparent. First, they are both examples of what are called **quasilinear** partial differential equations. That is to say, although their orders are different, in each case the *highest order derivatives occur only to degree one.* Thus (1) is a second order quasilinear equation and (2), (3) form a simultaneous first order quasilinear *system* of equations. Secondly, it is always possible to express a higher order equation as a first order system. Such a reduction is not unique but this turns out to be unimportant in most cases. To see one way in which this may be achieved for equation (1) set $u = \partial y/\partial t$, $v = \partial y/\partial x$, when (1) becomes

$$\frac{\partial u}{\partial t} = c^2(1 + v^2)^{-2}\frac{\partial v}{\partial x}. \tag{5}$$

To obtain the necessary second equation relating u and v we only need to use the fact that as the second order derivatives are assumed to exist and to be continuous we must have equality of mixed derivatives, giving

$$\frac{\partial v}{\partial t} = \frac{\partial u}{\partial x}. \tag{6}$$

System (5), (6) of two first order quasilinear equations now replaces the original second order equation. Although we shall not prove it here, it is a simple matter to check that solutions of (5), (6) are also solutions of (1).

It is most convenient to employ matrix notation when discussing both linear and quasilinear systems. Let us illustrate this first in the case of equations (5), (6). Define

$$\mathbf{A} = \begin{bmatrix} 0 & -c^2(1+v^2)^{-2} \\ -1 & 0 \end{bmatrix} \quad \text{and} \quad \mathbf{U} = \begin{bmatrix} u \\ v \end{bmatrix}, \tag{7}$$

then (5), (6) can be written as the *matrix equation*

$$\frac{\partial \mathbf{U}}{\partial t} + \mathbf{A} \frac{\partial \mathbf{U}}{\partial x} = 0, \tag{8}$$

with

$$\frac{\partial \mathbf{U}}{\partial t} = \begin{bmatrix} \dfrac{\partial u}{\partial t} \\ \dfrac{\partial v}{\partial t} \end{bmatrix} \quad \text{and} \quad \frac{\partial \mathbf{U}}{\partial x} = \begin{bmatrix} \dfrac{\partial u}{\partial x} \\ \dfrac{\partial v}{\partial x} \end{bmatrix}. \tag{9}$$

In the case of the system of equations (2) and (3), once the indicated differentiations have been performed, it is easily shown that in the one-dimensional case the matrix form of this system is

$$\frac{\partial \mathbf{U}}{\partial t} + \mathbf{A} \frac{\partial \mathbf{U}}{\partial x} = 0, \tag{10}$$

where

$$\mathbf{A} = \begin{bmatrix} u & \rho \\ a^2/\rho & u \end{bmatrix} \quad \text{and} \quad \mathbf{U} = \begin{bmatrix} \rho \\ u \end{bmatrix}. \tag{11}$$

Here we have used the result $p = f(\rho)$ to write, in one space dimension,

$$\operatorname{grad} p = \frac{\mathrm{d}p}{\mathrm{d}\rho} \frac{\partial \rho}{\partial x} = a^2 \frac{\partial \rho}{\partial x}, \tag{12}$$

where $a^2 = \mathrm{d}p/\mathrm{d}\rho$ is the square of the local sound speed in the gas.

§101   Conservation equation

In many physical problems the equations that arise come directly from the laws of conservation of some quantity, such as mass, momentum, energy or electric charge. Such laws are called **conservation laws**, and they express the balance between the rate of outflow of a quantity from a volume V and the time rate of change of the amount of that quantity that is contained within V. For a scalar quantity q such a conservation law has the general form

$$\frac{\partial q}{\partial t} + \operatorname{div} \mathbf{h}(q) = 0, \tag{13}$$

where $\mathbf{h}(q)$ is some vector function (linear or nonlinear) of q.

In one space dimension a *matrix conservation law* may be written

$$\frac{\partial \mathbf{U}}{\partial t} + \frac{\partial \mathbf{F}(\mathbf{U})}{\partial x} = \mathbf{0}, \tag{14}$$

with $\mathbf{U}$ and $\mathbf{F}(\mathbf{U})$ both column vectors. When the equations from which (14) are derived are linear constant coefficient equations it is always possible to write $\mathbf{F}(\mathbf{U}) = \mathbf{A}\mathbf{U}$, with $\mathbf{A}$ a constant coefficient square matrix. If the equations giving rise to (14) are quasilinear then column vector $\mathbf{F}$ depends nonlinearly on the elements of $\mathbf{U}$ through its own elements.

The situation is well illustrated by the one-dimensional form of the two equations (2), (3) which, after a little manipulation, can be written as conservation laws. Equation (2) is already in the precise form (13), since in one dimension it becomes

$$\frac{\partial \rho}{\partial t} + \frac{\partial}{\partial x}(\rho u) = 0.$$

Equation (3) can be put into this same form if it is first multiplied by ρ and then added to u times equation (2), for it may then be written

$$\frac{\partial}{\partial t}(\rho u) + \frac{\partial}{\partial x}(\rho u^2 + p) = 0, \tag{15}$$

which merely expresses the conservation of momentum. So the *matrix conservation law* expressing (2) and (15) is

$$\frac{\partial \mathbf{U}}{\partial t} + \frac{\partial \mathbf{F}}{\partial x} = \mathbf{0}, \tag{16}$$

with

$$\mathbf{U} = \begin{bmatrix} \rho \\ \rho u \end{bmatrix} \quad \text{and} \quad \mathbf{F} = \begin{bmatrix} \rho u \\ \rho u^2 + p \end{bmatrix}. \tag{17}$$

The nonlinear dependence of the elements of $\mathbf{F}$ on those of $\mathbf{U}$ is now clearly apparent.

§102   General effect of nonlinearity

It is now necessary to make clear that the effect of nonlinearity in a wave equation involves more than the loss of superposibility, for it can also change the entire nature of the solution. In the first instance this is best shown by a simple non-physical example.

Consider the single first order partial differential equation

$$\frac{\partial u}{\partial t} + f(u)\frac{\partial u}{\partial x} = 0 \tag{18}$$

for the scalar $u(x, t)$ that is subject to the *initial condition*

$$u(x, 0) = g(x). \tag{19}$$

Now the total differential du is given by

$$du = \frac{\partial u}{\partial t}dt + \frac{\partial u}{\partial x}dx,$$

so that if x and t are constrained to lie on a curve C, then at any point P on C we have

$$\frac{du}{dt} = \frac{\partial u}{\partial t} + \left(\frac{dx}{dt}\right)\frac{\partial u}{\partial x}, \tag{20}$$

where now dx/dt is the gradient of curve C at point P.

Comparison of (18) and (20) now shows that we may interpret (18) as the ordinary differential equation

$$\frac{du}{dt} = 0 \tag{21}$$

along any member of the family of curves C which are the solution curves of

$$\frac{dx}{dt} = f(u). \tag{22}$$

These curves C are called the **characteristic curves** of equation (18). The solution of the partial differential equation (18) has thus been reduced to the solution of the pair of *simultaneous ordinary differential equations* (21) and (22).

Equation (21) shows that $u = $ const. along each of the characteristic curves C. The constant value actually associated with any characteristic curve being equal to the value of u determined by the initial data (19) at the point at which the characteristic curve intersects the initial line $t = 0$. Setting $u = $ const. in (22) then shows that the characteristic curves C of (18) form a *family of straight lines*. So, if we consider the characteristic through the point $(\xi, 0)$ on the initial line, we find after integrating (22) and using (19) that the family of characteristic curves C have the equation

$$x = \xi + tf(g(\xi)), \tag{23}$$

where ξ now plays the role of a parameter.

Expressed slightly differently, we have shown that in terms of the parameter ξ, $u = g(\xi)$ at every point of the line (23) in the (x, t) plane. In physical problems t usually denotes time, so that it is then necessary to confine attention to the upper half plane in which $t \geqslant 0$.

The solution to (18) and (19) may be found in implicit form if ξ is eliminated between $u = g(\xi)$, which is true along a characteristic, and the equation (23) of the characteristic itself. We find the general result

$$u = g(x - tf(u)). \tag{24}$$

Result (23) is probably more instructive than (24), because it shows that if the functions f and g are such that two characteristics intersect for $t > 0$, then since each one will have associated with it a different constant value of u, it must follow that at such a point the solution will not be unique. This can obviously happen however smooth the two functions may be, since intersection of two characteristics depends merely on the value of $f(g(\xi))$ that is associated with each of the straight line characteristics. This is to say on the two points $(\xi_1, 0)$ and $(\xi_2, 0)$ of the initial line through which they pass. We conclude from this that such behaviour of solutions is not attributable to any irregularity in the coefficient $f(u)$, or in the initial data $u(x, 0) = g(x)$.

Behaviour of this nature has not been encountered elsewhere in this book and it typifies an important feature of quasilinear wave equations that we have not so far seen. Later we shall show that discontinuous solutions called *shocks* are possible in solutions of equations of this type. Their origin is attributable to precisely this sort of process, though a more complicated method of analysis than we can present here is required to show this for general quasilinear systems of wave equations.

This example may be developed a little further, because the family of characteristic curves C may have an *envelope* for $t > 0$, and since it is a family of straight lines this envelope is easy to find. If an envelope is formed then the envelope marks the locus of points at which non-uniqueness of the solution first starts and a shock forms (see §§**107** and **108**).

Now it is a familiar result from elementary calculus that if a family of curves C has the equation

$$\Phi(x, t, \xi) = 0, \tag{25}$$

with ξ a parameter, then the envelope, when it exists, is found by solving simultaneously (25) and the equation

$$\frac{\partial \Phi}{\partial \xi}(x, t, \xi) = 0. \tag{26}$$

Applying these results to (23) shows that, in terms of the parameter ξ, the envelope is defined by

$$x = \xi + tf(g(\xi)) \quad \text{and} \quad t = -1/\{f'(g(\xi))g'(\xi)\}. \tag{27}$$

When t is time we shall only be interested in the case when $t > 0$.

Differentiating (24) partially with respect to x gives

$$\frac{\partial u}{\partial x} = \frac{g'(x - tf(u))}{1 + tg'(x - tf(u))f'(u)}, \tag{28}$$

showing $\partial u / \partial x$ becomes infinite whenever $1 + tg'(x - tf(u))f'(u) = 0$. So, finally, inspection of (27) and (28) then makes it clear that $\partial u / \partial x$ will actually become infinite on the envelope of the characteristics. Hence, whenever an envelope exists, the solution will *steepen* as it is approached.

Many physical effects owe their existence to this form of nonlinear behaviour which has no counterpart in the linear theory of wave equations. Typical of these effects is the formation of a **bore** in a river. This occurs under certain conditions in a tidal river when the incoming tide gives rise to an almost vertical wall of water (a bore), which then propagates along the river in a remarkably stable manner for an appreciable distance.

§103 Characteristics

The notion of a characteristic curve introduced briefly in the previous section requires generalisation if it is to be applied to quasilinear systems of first order equations as typified by equations (10) and (11).

We now make this generalisation for the system

$$\frac{\partial \mathbf{U}}{\partial t} + \mathbf{A}\frac{\partial \mathbf{U}}{\partial x} + \mathbf{B} = 0, \tag{29}$$

in which $\mathbf{U}$ and $\mathbf{B}$ are n element column vectors with elements $u_1, u_2, \ldots, u_n$ and $b_1, b_2, \ldots, b_n$, respectively, and $\mathbf{A}$ is an $n \times n$ matrix with elements a_{ij}. The system (29) will be **quasilinear** if, in general, the elements a_{ij} of $\mathbf{A}$ depend nonlinearly on $u_1, u_2, \ldots, u_n$. When $\mathbf{B} \neq 0$ the elements b_i of $\mathbf{B}$ may, or may not, depend linearly on $u_1, u_2, \ldots, u_n$. It will be assumed throughout this section that the elements b_i and a_{ij} are *continuous functions* of their arguments.

It follows from what has been said so far that both of the simple systems discussed in §**100** are of the form (29), each with $n = 2$. They are, however, *homogeneous* since in each case $\mathbf{B} \equiv 0$.

Although x, t are the natural variables to use when deriving systems of equations describing motion in space and time, they are not necessarily the most appropriate ones from the mathematical point of view. So, as we are interested in the way a solution evolves with time, let us leave the time variable unchanged in system (29), but replace x by some arbitrary curvilinear coordinate ξ and then try to choose ξ in a manner which is convenient for our mathematical arguments. Accordingly, our starting point will be to change from (x, t) to the *arbitrary semi-curvilinear coordinates*

$$\xi = \xi(x, t), \qquad t' = t. \tag{30}$$

If the Jacobian of the transformation (30) is non-vanishing we may thus transform (29) by the rule

$$\frac{\partial}{\partial t} \equiv \frac{\partial \xi}{\partial t}\frac{\partial}{\partial \xi} + \frac{\partial t'}{\partial t}\frac{\partial}{\partial t'} \equiv \frac{\partial \xi}{\partial t}\frac{\partial}{\partial \xi} + \frac{\partial}{\partial t'}$$

$$\frac{\partial}{\partial x} \equiv \frac{\partial \xi}{\partial x}\frac{\partial}{\partial \xi} + \frac{\partial t'}{\partial x}\frac{\partial}{\partial t'} \equiv \frac{\partial \xi}{\partial x}\frac{\partial}{\partial \xi}$$

where, of course, $\partial \xi / \partial t$ and $\partial \xi / \partial x$ are scalar quantities. This leads directly to the transformed equation

$$\frac{\partial \mathbf{U}}{\partial t'} + \frac{\partial \xi}{\partial t}\frac{\partial \mathbf{U}}{\partial \xi} + \frac{\partial \xi}{\partial x}\mathbf{A}\frac{\partial \mathbf{U}}{\partial \xi} + \mathbf{B} = 0,$$

the terms of which may be grouped to yield

$$\frac{\partial \mathbf{U}}{\partial t'} + \left(\frac{\partial \xi}{\partial t}\mathbf{I} + \frac{\partial \xi}{\partial x}\mathbf{A}\right)\frac{\partial \mathbf{U}}{\partial \xi} + \mathbf{B} = \mathbf{0}, \qquad (31)$$

where $\mathbf{I}$ is the $n \times n$ unit matrix.

Equation (31) may now be considered to be an algebraic relationship connecting the matrix vector derivatives $\partial \mathbf{U}/\partial t'$ and $\partial \mathbf{U}/\partial \xi$. It is then at once apparent that this equation may only be used to determine $\partial \mathbf{U}/\partial \xi$ if the inverse of the coefficient matrix of $\partial \mathbf{U}/\partial \xi$ exists. That is to say, if the determinant of the coefficient matrix of $\partial \mathbf{U}/\partial \xi$ is non-vanishing. This condition obviously depends on the nature of the curvilinear coordinate lines $\xi(x, t) = \text{const.}$, which so far have been chosen arbitrarily. Suppose now that for the particular choice $\xi \equiv \varphi$ the determinant does vanish, giving the condition

$$\left|\frac{\partial \varphi}{\partial t}\mathbf{I} + \frac{\partial \varphi}{\partial x}\mathbf{A}\right| = 0. \qquad (32)$$

Then because of this the derivative $\partial \mathbf{U}/\partial \varphi$ will be indeterminate on the family of lines $\varphi = \text{const.}$ Consequently, across such lines $\varphi(x, t) = \text{const.}$, $\partial \mathbf{U}/\partial \varphi$ may actually be discontinuous. This means that each of the n elements $\partial u_i/\partial \varphi$ of $\partial \mathbf{U}/\partial \varphi$ may be discontinuous across any of the lines $\varphi = \text{const.}$ To find how, when they occur, these discontinuities in $\partial u_i/\partial \varphi$ are related one to the other across a curvilinear coordinate line $\varphi = \text{const.}$, it is necessary to reconsider equation (31).

We shall now confine attention to solutions $\mathbf{U}$ which are *everywhere continuous* but for which the *derivative* $\partial \mathbf{U}/\partial \varphi$ is *discontinuous* across the particular line $\varphi = k$ (say). Because of the continuity of $\mathbf{U}$, and the continuity of the elements a_{ij} of $\mathbf{A}$ and b_i of $\mathbf{B}$, the matrices $\mathbf{A}$ and $\mathbf{B}$ will experience no discontinuity across $\varphi = k$. So, in the neighbourhood of a typical point P of this line, $\mathbf{A}$ and $\mathbf{B}$ may be given their actual values *at P*. In equation (31) there is no indeterminacy of $\partial \mathbf{U}/\partial t'$ across the lines $\varphi = \text{const.}$, and as $\partial/\partial t'$ denotes differentiation *along* these lines it must follow that $\partial \mathbf{U}/\partial t'$ is everywhere continuous and, in particular, that it is *continuous across the line $\varphi = k$ at P*.

Taking these facts into account and differencing equation (31) across the line $\xi \equiv \varphi = k$ at P gives

$$\left(\frac{\partial \varphi}{\partial t}\mathbf{I} + \frac{\partial \varphi}{\partial x}\mathbf{A}\right)_P \left[\!\left[\frac{\partial \mathbf{U}}{\partial \varphi}\right]\!\right]_P = \mathbf{0}, \qquad (33)$$

where $[\![\alpha]\!] \equiv \alpha_- - \alpha_+$ signifies the discontinuous jump in the quantity α across the line $\varphi = k$, with α_- denoting the value to the immediate left of the line and α_+ the value to the immediate right at P. As the point P was any point on this line the suffix P may now be omitted. The operator $\partial/\partial\varphi$ is *differentiation normal to the curves* $\varphi = $ const., so that equations (33) express compatibility conditions to be satisfied by the component of the derivative of $\mathbf{U}$ on either side of and normal to these curves in the (x, t)-plane.

This is a homogeneous system of equations for the n jump quantities $[\![\partial u_i/\partial\varphi]\!] \equiv (\partial u_i/\partial\varphi)_- - (\partial u_i/\partial\varphi)_+$ and there will only be a non-trivial solution if the determinant of the coefficients vanishes. The condition for this is

$$\left| \frac{\partial\varphi}{\partial t}\mathbf{I} + \frac{\partial\varphi}{\partial x}\mathbf{A} \right| = 0. \tag{34}$$

However, along the lines $\varphi = $ const. we have, by differentiation,

$$\frac{\partial\varphi}{\partial t} + \frac{\partial\varphi}{\partial x}\frac{dx}{dt} = 0,$$

so that these lines have the gradient

$$\frac{dx}{dt} = -\frac{\partial\varphi}{\partial t} \bigg/ \frac{\partial\varphi}{\partial x} \equiv \lambda \text{ (say)}. \tag{35}$$

Combining (34) and (35) we deduce that λ must be such that

$$|\mathbf{A} - \lambda\mathbf{I}| = 0. \tag{36}$$

Consequently the λ in (35) can only be one of the eigenvalues of $\mathbf{A}$, and since (33) can be re-written

$$(\mathbf{A} - \lambda\mathbf{I})\left[\!\!\left[\frac{\partial\mathbf{U}}{\partial\varphi}\right]\!\!\right] = \mathbf{0}, \tag{37}$$

the column vector $[\![\partial\mathbf{U}/\partial\varphi]\!]$ must be proportional to the corresponding right eigenvector of $\mathbf{A}$. This, then, determines the ratios between the n elements $[\![\partial u_i/\partial\varphi]\!]$ of the vector $[\![\partial\mathbf{U}/\partial\varphi]\!]$ that we were seeking.

As $\mathbf{A}$ is an $n \times n$ matrix it will have n eigenvalues. If these are real and distinct, integration of equations (35) will give rise to n distinct families of real curves $C^{(1)}, C^{(2)}, \ldots, C^{(n)}$ in the (x, t) plane:

$$C^{(i)}: \frac{dx}{dt} = \lambda^{(i)}, \qquad i = 1, 2, \ldots, n. \tag{38}$$

If x denotes a distance and t a time, the eigenvalues will have the dimensions of a *speed*. Any one of these families of curves $C^{(i)}$ may be taken for our curvilinear coordinate lines $\varphi = $const. The $\lambda^{(i)}$ associated with each family will then be the speed of propagation of the matrix column vector $[\![\partial U/\partial\varphi]\!]$ along the curves $C^{(i)}$ belonging to that family.

When the eigenvalues $\lambda^{(i)}$ of A are all real and distinct, so that the propagation speeds are also all real and distinct, and there are n distinct linearly independent right eigenvectors $r^{(i)}$ of A satisfying the defining relation

$$Ar^{(i)} = \lambda^{(i)}r^{(i)}, \quad \text{for } i = 1, 2, \ldots, n, \tag{39}$$

the system of equations (29) will be said to be **totally hyperbolic**. We may, if we desire, replace the words right eigenvector by left eigenvector in this definition, where the left eigenvectors l of A satisfy the defining relation

$$l^{(i)}A = \lambda^{(i)}l^{(i)}, \quad \text{for } i = 1, 2, \ldots, n. \tag{40}$$

This follows because simple linear algebra arguments establish that when n linearly independent vectors $r^{(i)}$ exist, then so also do n linearly independent vectors $l^{(i)}$.

Hereafter our concern will be with such systems, since they characterise the type of wave propagation that has been the object of our study so far. The families of curves $C^{(i)}$ defined by integration of equations (38) are called the families of **characteristic curves** of system (29). A totally hyperbolic system (29) is thus one in which there are n distinct real speeds of propagation of a disturbance, each of which when characterised by the appropriate right eigenvector is different. The precise nature of these differences will be examined shortly.

The relationship between characteristic curves and the solution vector U to system (29) is illustrated in Fig. 37 in the case of a typical element u_i of U. Here it has been assumed that initial conditions have been specified for system (29) in the form

$$U(x, 0) = \Psi(x),$$

where the ith element u_i of U has for its initial condition $u_i(x, 0) = \psi_i(x)$.

The line PQ in the solution surface S is the one across which $\partial u_i/\partial\varphi$ is discontinuous, and its projection onto the (x, t) plane is the characteristic which has equation $\varphi(x, t) = k$. Since such a line marks the boundary between the different solutions to the left and right of it, it is natural to think of the solution to the left of $\varphi = k$ as a *propagating*

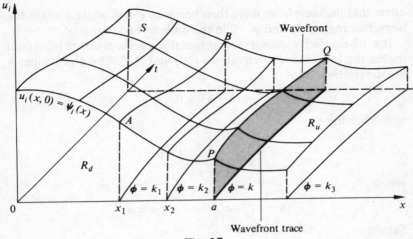

Fig. 37

disturbance wave, and the solution to the right as the solution in a region not yet reached by the disturbance. With these ideas in mind we shall call the line PQ in S the *solution surface* **wavefront**, its projection onto the (x, t) plane which forms the characteristic curve $\varphi = k$ the **wavefront trace**, the region R_d to the left of $\varphi = k$ the **disturbed region** and the region R_u to the right of $\varphi = k$ the **undisturbed region**. At any time t_1 the *physical* **wavefront** is at the intersection of the wavefront trace and the line $t = t_1$.

Since it was not necessary that $\partial \mathbf{U}/\partial \varphi$ should be discontinuous across the characteristics $\varphi = $ const., it must follow that continuous and differentiable elements of the initial data $u_i(x, 0) = \psi_i(x)$ will also propagate along characteristics. In the case of the element of initial data at A, this will propagate along the characteristic $\varphi = k_1$ (say) starting from the point $(x_1, 0)$ which is the projection of A onto the initial line. The characteristic $\varphi = k_1$ is then the projection onto the (x, t) plane of the path AB followed by the element of the solution surface S that started at A. Characteristics corresponding to $k = k_2, k_3, k_4$ etc., may be interpreted in similar fashion.

The way the constants $k_1, k_2, \ldots$ are assigned to different characteristics is arbitrary, apart from the fact that $\varphi = $ const. are coordinate lines so that k must be assigned *monotonically* along the initial line $t = 0$. If it is necessary to do this, then possibly the simplest and most convenient method of parameterisation is to assign to the characteristic through $(x, 0)$ on the initial line the value $\varphi = x - a$. This has the effect that the wavefront trace then becomes $\varphi = 0$, while $\varphi < 0$ in the disturbed region R_d and $\varphi > 0$ in the undisturbed region R_u.

It will help clarify ideas if we consider the specific problem from fluid mechanics introduced in equations (10) and (11). The eigenvalues λ are determined by

$$|\mathbf{A} - \lambda \mathbf{I}| = 0,$$

which becomes

$$\begin{vmatrix} u - \lambda & \rho \\ a^2/\rho & u - \lambda \end{vmatrix} = 0,$$

giving

$$(u - \lambda)^2 = a^2 \quad \text{or} \quad \lambda = u \pm a.$$

Setting

$$\lambda^{(1)} = u + a, \qquad \lambda^{(2)} = u - a, \tag{41}$$

the families of characteristic curves $C^{(1)}$ and $C^{(2)}$ defined by (38) become

$$C^{(1)}: \frac{\mathrm{d}x}{\mathrm{d}t} = u + a \quad \text{and} \quad C^{(2)}: \frac{\mathrm{d}x}{\mathrm{d}t} = u - a. \tag{42}$$

Expressed in physical terms we see that in the $C^{(1)}$ family, disturbances thus propagate with the sum of the fluid speed and the sound speed. In the $C^{(2)}$ family they propagate with the difference of the fluid speed and the sound speed.

The right eigenvectors $\mathbf{r}^{(1)}$ and $\mathbf{r}^{(2)}$ must, by (39), satisfy

$$(\mathbf{A} - \lambda^{(i)}\mathbf{I})\mathbf{r}^{(i)} = \mathbf{0}. \qquad (i = 1, 2) \tag{43}$$

So, denoting the elements of $\mathbf{r}^{(i)}$ by $r_1^{(i)}$ and $r_2^{(i)}$, we have from (11), (41) and (43)

$$\begin{bmatrix} u - \lambda^{(i)} & \rho \\ a^2/\rho & u - \lambda^{(i)} \end{bmatrix} \begin{bmatrix} r_1^{(i)} \\ r_2^{(i)} \end{bmatrix} = \mathbf{0}, \, i = 1, 2. \tag{44}$$

Since only the *ratios* of the elements of the right eigenvectors are determined by (43) it is often convenient to normalise the eigenvectors so that the first element is unity, when it follows from (44) that

$$\mathbf{r}^{(1)} = \begin{bmatrix} 1 \\ a/\rho \end{bmatrix} \quad \text{and} \quad \mathbf{r}^{(2)} = \begin{bmatrix} 1 \\ -a/\rho \end{bmatrix}. \tag{45}$$

It may thus be concluded that wherever a wavefront exists, since the form of $\mathbf{U}$ in (11) implies

$$\left[\!\!\left[\frac{\partial \mathbf{U}}{\partial \varphi}\right]\!\!\right] = \begin{bmatrix} [\![\partial\rho/\partial\varphi]\!] \\ [\![\partial u/\partial\varphi]\!] \end{bmatrix},$$

a comparison of (37) and (43) followed by the use of (45) will yield:

across a wavefront in the $C^{(1)}$ family of characteristic curves

$$\frac{[\![\partial\rho/\partial\varphi]\!]}{1} = \frac{[\![\partial u/\partial\varphi]\!]}{a/\rho}, \tag{46}$$

and

across a wavefront in the $C^{(2)}$ family of characteristic curves

$$\frac{[\![\partial\rho/\partial\varphi]\!]}{1} = \frac{[\![\partial u/\partial\varphi]\!]}{-a/\rho}, \tag{47}$$

where a and ρ in (46) and (47) have values appropriate to the wavefront.

This result generalises without difficulty, for suppose that the vector $\mathbf{r}^{(i)}$ with elements $r_1^{(i)}, r_2^{(i)}, \ldots, r_n^{(i)}$ is the ith right eigenvector of $\mathbf{A}$ corresponding to the eigenvalue $\lambda = \lambda^{(i)}$. Then across a wavefront belonging to the $C^{(i)}$ family we may write,

$$\frac{[\![\partial u_1/\partial\varphi]\!]}{r_1^{(i)}} = \frac{[\![\partial u_2/\partial\varphi]\!]}{r_2^{(i)}} = \ldots = \frac{[\![\partial u_n/\partial\varphi]\!]}{r_n^{(i)}}, \tag{48}$$

where the elements of $\mathbf{r}^{(i)} = \mathbf{r}^{(i)}(\mathbf{U})$ have values determined by $\mathbf{U}$ on the wavefront. This is a result we shall have occasion to use again later.

§104   Wavefronts bounding a constant state

In physical situations the solution vector $\mathbf{U}$ describes the "state" of the system described by equations (29). It is thus convenient to refer to a region in which $\mathbf{U}$ is non-constant as a **disturbed state**, and a region in which $\mathbf{U}$ is constant as a **constant state**, irrespective of whether or not the system involved described a physical situation. Our purpose here will be to examine the simplification that results in equations like (46) and (47) when a wavefront bounds a constant state.

First, as the elements a_{ij} of $\mathbf{A}$ are continuous functions of their arguments, it follows directly that the eigenvalues $\lambda^{(i)}$ of $\mathbf{A}$ are continuous functions of a_{ij}, and hence of the elements $u_1, u_2, \ldots, u_n$ of **U**. Since **U** is itself continuous across a wavefront we conclude that $\lambda^{(i)} = \lambda_0^{(i)} = \text{const.}$, on a wavefront bounding the constant state $\mathbf{U} = \mathbf{U}_0$, where $\lambda_0^{(i)} = \lambda^{(i)}(\mathbf{U}_0)$. From equations (38) we thus see that if a characteristic curve from the ith family $C^{(i)}$ bounds a constant state, then it must be a straight line.

If such a straight line characteristic $C_0^{(i)}$ belonging to the ith family $C^{(i)}$ bounds a constant state $\mathbf{U} = \mathbf{U}_0$ that lies to its right (say), then because $(\partial \mathbf{U}/\partial \varphi)_+ = \partial \mathbf{U}_0/\partial \varphi \equiv \mathbf{0}$,

$$\left[\!\left[\frac{\partial u_j}{\partial \varphi}\right]\!\right] \equiv \left(\frac{\partial u_j}{\partial \varphi}\right)_- - \left(\frac{\partial u_j}{\partial \varphi}\right)_+ = \left(\frac{\partial u_j}{\partial \varphi}\right)_- \quad \text{for } j = 1, 2, \ldots, n. \tag{49}$$

Now $\partial \mathbf{U}/\partial t'$ is continuous across $C_0^{(i)}$ while $\partial \mathbf{U}_0/\partial t' \equiv \mathbf{0}$. Thus in the disturbed region immediately adjacent to $C_0^{(i)}$ the total differential $\mathrm{d}u_j$ reduces to

$$\mathrm{d}u_j = \left(\frac{\partial u_j}{\partial \varphi}\right)_- \mathrm{d}\varphi \quad \text{for } j = 1, 2, \ldots, n. \tag{50}$$

By virtue of (48) and (49) this is equivalent to

$$\mathrm{d}u_j = K r_j^{(i)} \mathrm{d}\varphi, \tag{51}$$

where K is some constant of proportionality. It proves convenient to choose K so that the first element $K r_1^{(i)}$ of $K\mathbf{r}^{(i)}$ becomes unity. Setting $j = 1$ in (51) then gives $\mathrm{d}u_1 = \mathrm{d}\varphi$, so that all the other differentials $\mathrm{d}u_2, \mathrm{d}u_3, \ldots, \mathrm{d}u_n$ become expressible in terms of $\mathrm{d}u_1$, because (51) becomes

$$\mathrm{d}u_j = r_j^{(i)} \mathrm{d}u_1 \quad \text{for } j = 1, 2, \ldots, n \quad \text{or} \quad \mathrm{d}\mathbf{U} = \mathbf{r}^{(i)} \mathrm{d}u_1. \tag{52}$$

The convenience of this normalisation is well illustrated by means of the eigenvectors given in (45) for the equations of one-dimensional unsteady isentropic gas flow. We find that in the disturbed region immediately adjacent to a wavefront belonging to the $C^{(1)}$ family of characteristics that bounds a constant state u_0, ρ_0 (in which $a = a_0$), it must follow from $\mathbf{r}^{(1)}$ that

$$\mathrm{d}u = (a_0/\rho_0) \, \mathrm{d}\rho. \tag{53}$$

The corresponding behaviour adjacent to a wavefront belonging to the $C^{(2)}$ family that bounds this same constant state is, from $\mathbf{r}^{(2)}$,

$$\mathrm{d}u = -(a_0/\rho_0) \, \mathrm{d}\rho. \tag{54}$$

A simple rule that is sometimes useful for deriving results of this form follows by combining the matrix vector form of (52) and the defining relationship

$$\mathbf{Ar} = \lambda \mathbf{r}$$

for the right eigenvector corresponding to the eigenvalue λ. Immediately adjacent to the constant state $\mathbf{U} = \mathbf{U}_0$ this gives the result

$$(\mathbf{A}_0 - \lambda_0 \mathbf{I}) \, d\mathbf{U} = 0, \tag{55}$$

where $\mathbf{A}_0 = \mathbf{A}(\mathbf{U}_0)$ and $\lambda_0 = \lambda(\mathbf{U}_0)$. Comparison of this result with system (29) from which it was derived now yields the following rule.

Rule for compatibility conditions for elements of dU. To find the relationships that exist between the elements $du_1, du_2, \ldots, du_n$ of $d\mathbf{U}$ in the disturbed region immediately adjacent to a wavefront that bounds a constant state $\mathbf{U} = \mathbf{U}_0$, the vector $\mathbf{B}$ in (29) should be neglected, the undifferentiated variables should be replaced by their constant state values, and in the differentiated terms the following replacements should be made

$$\frac{\partial}{\partial t} \to -\lambda \, d(\,.\,) \quad \text{and} \quad \frac{\partial}{\partial x} \to d(\,.\,). \tag{56}$$

For example, a term $\partial \rho / \partial t$ should be replaced by $-\lambda \, d\rho$ and one like $u \, \partial u / \partial x$ should be replaced by $u_0 \, du$. When applied to the scalar equations represented by (10) and (11) which gave rise to (53) and (54) this rule gives

$$-\lambda \, d\rho + u_0 \, d\rho + \rho_0 \, du = 0,$$

$$-\lambda \, du + u_0 \, du + \frac{a_0^2}{\rho_0} d\rho = 0.$$

Then, as before, we find

$$\lambda_0^{(1)} = u_0 + a_0, \qquad \lambda_0^{(2)} = u_0 - a_0,$$

while for a $C^{(1)}$ characteristic wavefront $du = (a_0/\rho_0) \, d\rho$ and for a $C^{(2)}$ characteristic wavefront $du = -(a_0/\rho_0) \, d\rho$.

§105   Riemann invariants

In this section we offer a brief discussion of an important technique that can lead directly to a solution when certain calculations can be performed, and which in any case provides a valuable insight into the nature of solutions to a special class of problems. This method applies to any totally hyperbolic system of two homogeneous first order equations involving two dependent variables u_1, u_2 of the general form

$$\frac{\partial u_1}{\partial t} + a_{11}\frac{\partial u_1}{\partial x} + a_{12}\frac{\partial u_2}{\partial x} = 0,$$

$$\frac{\partial u_2}{\partial t} + a_{21}\frac{\partial u_1}{\partial x} + a_{22}\frac{\partial u_2}{\partial x} = 0, \qquad (57)$$

which is subject to the initial data

$$u_1(x, 0) = \tilde{u}_1(x) \quad \text{and} \quad u_2(x, 0) = \tilde{u}_2(x). \qquad (58)$$

The coefficients $a_{ij} = a_{ij}(u_1, u_2)$ will, in general, be assumed to be functions of the two dependent variables u_1 and u_2, but not to have any explicit dependence on the independent variables x and t. The system (57) will be **quasilinear** when $a_{ij} = a_{ij}(u_1, u_2)$ and it will be **linear** in the special case when the coefficients a_{ij} are all constants.

Defining **A** and **U** to be

$$\mathbf{A} = \begin{bmatrix} a_{11} & a_{12} \\ a_{21} & a_{22} \end{bmatrix}, \qquad \mathbf{U} = \begin{bmatrix} u_1 \\ u_2 \end{bmatrix}$$

enables equations (57) to be written

$$\frac{\partial \mathbf{U}}{\partial t} + \mathbf{A}\frac{\partial \mathbf{U}}{\partial x} = \mathbf{0}, \qquad (59)$$

when we know from §103 that the system will be *totally hyperbolic* provided the two eigenvalues $\lambda^{(i)}$, $i = 1, 2$ of

$$|\mathbf{A} - \lambda \mathbf{I}| = 0 \qquad (60)$$

are real and distinct and **A** has two linearly independent eigenvectors. In place of the right eigenvectors **r** that were useful in §103, and which were defined in (39), let us now make use of the corresponding left eigenvectors **l** defined in (40) by

$$\mathbf{l}^{(i)}\mathbf{A} = \lambda^{(i)}\mathbf{l}^{(i)}, \quad \text{for } i = 1, 2. \qquad (61)$$

If, now, we pre-multiply (59) by $\mathbf{l}^{(i)}$ and use (61) we obtain the result

$$\mathbf{l}^{(i)}\left(\frac{\partial \mathbf{U}}{\partial t} + \lambda^{(i)}\frac{\partial \mathbf{U}}{\partial x}\right) = 0, \quad \text{for } i = 1, 2. \qquad (62)$$

In this the bracketed expression will be recognised as the directional derivative of **U** with respect to time along the family of characteristics $C^{(i)}$. Denoting differentiation with respect to time along members of the $C^{(1)}$ family of characteristics by $d/d\alpha$ and differentiation with respect to time along members of the $C^{(2)}$ family of characteristics by $d/d\beta$ enables us to replace (62) by the following pair of ordinary differential equations which are defined

along the $C^{(1)}$ characteristics by

$$\mathbf{l}^{(1)}\frac{d\mathbf{U}}{d\alpha}=0, \tag{63}$$

and along the $C^{(2)}$ characteristics by

$$\mathbf{l}^{(2)}\frac{d\mathbf{U}}{d\beta}=0. \tag{64}$$

Hence $\beta = $ const., along $C^{(1)}$ characteristics and $\alpha = $ const., along $C^{(2)}$ characteristics as indicated in Fig. 38.

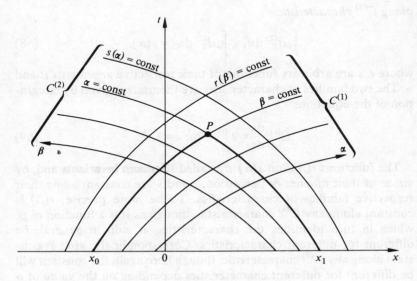

Fig. 38

Setting the left eigenvector $\mathbf{l}^{(i)} = (l_1^{(i)}, l_2^{(i)})$, for $i = 1, 2$ then enables (63), (64) to be re-expressed as

$$l_1^{(1)}\frac{\mathrm{d}u_1}{\mathrm{d}\alpha} + l_2^{(1)}\frac{\mathrm{d}u_2}{\mathrm{d}\alpha} = 0 \quad \text{along the } C^{(1)} \text{ characteristics} \tag{65}$$

and

$$l_1^{(2)}\frac{\mathrm{d}u_1}{\mathrm{d}\beta} + l_2^{(2)}\frac{\mathrm{d}u_2}{\mathrm{d}\beta} = 0 \quad \text{along the } C^{(2)} \text{ characteristics.} \tag{66}$$

Since, by supposition, $\mathbf{A}$ depends only on u_1 and u_2, so also will the coefficients $l_j^{(i)}$ of the left eigenvectors $\mathbf{l}^{(1)}$, $\mathbf{l}^{(2)}$. Consequently, both (65) and (66) will always be integrable along their respective characteristics, though they may first require multiplication by a suitable integrating factor μ.

Integrating (65) with respect to α along the $C^{(1)}$ characteristics, and (66) with respect to β along the $C^{(2)}$ characteristics gives:

along $C^{(1)}$ characteristics

$$\int \mu l_1^{(1)} \,\mathrm{d}u_1 + \int \mu l_2^{(1)} \,\mathrm{d}u_2 = r(\beta) \tag{67}$$

and

along $C^{(2)}$ characteristics

$$\int \mu l_1^{(2)} \,\mathrm{d}u_1 + \int \mu l_2^{(2)} \,\mathrm{d}u_2 = s(\alpha), \tag{68}$$

where r, s are arbitrary functions of their respective arguments β and α. The two families of characteristics are themselves given by integration of the equations

$$C^{(i)}: \frac{\mathrm{d}x}{\mathrm{d}t} = \lambda^{(i)}, \quad \text{for } i = 1, 2. \tag{69}$$

The functions $r(\beta)$ and $s(\alpha)$ are called **Riemann invariants** and, by virtue of their manner of derivation, r and s are constant along their respective families of characteristics. To be more precise, $r(\beta)$ is constant along any $C^{(1)}$ characteristic, though as it is a function of β, which in turn identifies the characteristics, it will, in general, be different for different characteristics. Correspondingly, $s(\alpha)$ is constant along any $C^{(2)}$ characteristic, though here again the constant will be different for different characteristics depending on the value of α associated with each characteristic.

Equations (67) and (68) enable u_1 and u_2 to be expressed in terms of r and s, the values of which are determined at points of the initial line $t = 0$ by the initial data (58). Suppose $r(\beta)$ in (67) is denoted by $R(u_1, u_2)$ and $s(\alpha)$ in (68) is denoted by $S(u_1, u_2)$. Then along the $C^{(1)}$ characteristic issuing out from the point $(x_0, 0)$ of the initial line in the sense of increasing time we have from (58) and the property of $r(\beta)$ that

$$R(u_1, u_2) = R(\tilde{u}_1(x_0), \tilde{u}_2(x_0)). \qquad (70)$$

Similarly, along the $C^{(2)}$ characteristic issuing out from the point $(x_1, 0)$ of the initial line in the sense of increasing time we have from (58) and the property of $s(\alpha)$ that

$$S(u_1, u_2) = S(\tilde{u}_1(x_1), \tilde{u}_2(x_1)). \qquad (71)$$

Solving these two implicit equations for u_1 and u_2 then determines the solution at the point P in Fig. 38 which is the point of intersection of the $C^{(1)}$ and $C^{(2)}$ characteristics along which the respective constant values of R and S are transported. In principle the initial value problem is now solved, since as the points $(x_0, 0)$ and $(x_1, 0)$ of the initial line were arbitrary, so also is the point P which may be anywhere in the upper half plane. However, in any particular case, the task of solving the two implicit relationships and of finding the characteristic curves in order to determine their point of intersection P is usually difficult. Nevertheless, this method of solution can often be used to solve problems and it is, in any case, of considerable theoretical importance.

To illustrate the method we apply it to the system of equations (10), (11) which are of the precise form given in (57). The eigenvalues of this system have already been found in (41) and the characteristic curves follow by integration of (42). A simple calculation shows that the left eigenvector $\mathbf{l}^{(1)}$ corresponding to $\lambda^{(1)} = u + a$ is

$$\mathbf{l}^{(1)} = (1, \rho/a) \qquad (72)$$

and the left eigenvector corresponding to $\lambda^{(2)} = u - a$ is

$$\mathbf{l}^{(2)} = (1, -\rho/a). \qquad (73)$$

Making the identifications $u_1 = \rho$, $u_2 = u$ equations (65), (66) then become

$$\frac{d\rho}{d\alpha} + \frac{\rho}{a}\frac{du}{d\alpha} = 0 \quad \text{along } C^{(1)} \text{ characteristics} \qquad (74a)$$

and

$$\frac{d\rho}{d\beta} - \frac{\rho}{a} \frac{du}{d\beta} = 0 \quad \text{along } C^{(2)} \text{ characteristics.} \tag{74b}$$

As $a = a(\rho)$, the integrating factor is seen by inspection to be $\mu = a/\rho$, so that (67) and (68) become

$$\int \frac{a}{\rho} d\rho + u = r(\beta) \quad \text{along } C^{(1)} \text{ characteristics} \tag{75}$$

and

$$\int \frac{a}{\rho} d\rho - u = s(\alpha) \quad \text{along } C^{(2)} \text{ characteristics.} \tag{76}$$

If the gas law $p = k\rho^\gamma$ is assumed then, since $a^2 = dp/d\rho$, it follows that (75) and (76) finally take the form

$$\frac{2a}{\gamma - 1} + u = r(\beta) \tag{77}$$

is constant along $C^{(1)}$ characteristics, and

$$\frac{2a}{\gamma - 1} - u = s(\alpha) \tag{78}$$

is constant along $C^{(2)}$ characteristics. Hence, working in terms of a rather than ρ, we find

$$a = (\gamma - 1)(r + s)/4, \quad u = (r - s)/2, \tag{79}$$

showing, as has already been remarked, that it is possible to express the dependent variables in terms of the Riemann invariants. The determination of the characteristics by integrating

$$C^{(1)}: \frac{dx}{dt} = u + a \quad \text{and} \quad C^{(2)}: \frac{dx}{dt} = u - a \tag{80}$$

is, however, only possible in special cases.

We shall look in more detail at an important special case of Riemann invariants in the next section, but in the meantime we conclude this section by making an application of the method to a linear problem. This both enables us to solve a complete problem and to illustrate the use of the method in the linear case.

Consider the pair of equations

$$\frac{\partial u_1}{\partial t} + \frac{\partial u_2}{\partial x} = 0,$$

$$\frac{\partial u_2}{\partial t} + \frac{\partial u_1}{\partial x} = 0,$$ (81)

where

$$u_1(x, 0) = e^x \quad \text{and} \quad u_2(x, 0) = e^{-x}.$$ (82)

Then the matrix $\mathbf{A}$ in (59) is

$$\mathbf{A} = \begin{bmatrix} 0 & 1 \\ 1 & 0 \end{bmatrix},$$

so that the eigenvalues and left eigenvectors are easily seen to be

$$\lambda^{(1)} = 1, \qquad \lambda^{(2)} = -1, \qquad \mathbf{l}^{(1)} = [1, 1], \qquad \mathbf{l}^{(2)} = [1, -1].$$

The problem is thus seen to be totally hyperbolic. When integrated, equations (65) and (66) give

$$u_1 + u_2 = r(\beta) \quad \text{along } C^{(1)} \text{ characteristics}$$

and

$$u_1 - u_2 = s(\alpha) \quad \text{along } C^{(2)} \text{ characteristics}.$$

The characteristics follow by integrating (69) to get

$$C^{(1)}: x = x_0 + t \quad \text{and} \quad C^{(2)}: x = x_1 - t,$$ (83)

with x_0 and x_1 arbitrary constants of integration.

Using the initial data (82) shows that at the point $(x_0, 0)$ of the initial line

$$u_1(x_0, 0) + u_2(x_0, 0) = e^{x_0} + e^{-x_0} = 2 \cosh x_0$$

while at the point $(x_1, 0)$ of the initial line

$$u_1(x_1 0) - u_2(x, 0) = e^{x_1} - e^{-x_1} = 2 \sinh x_1.$$

Hence it follows directly that the Riemann invariants are

$$u_1(x, t) + u_2(x, t) = 2 \cosh x_0$$

along the $C^{(1)}$ characteristic through the point $(x_0, 0)$ of the initial line and

$$u_1(x, t) - u_2(x, t) = 2 \sinh x_1$$

along the $C^{(2)}$ characteristic through the point $(x_1, 0)$ of the initial line.

Using the results $x_0 = x - t$, $x_1 = x + t$ that follow from the equations (83) describing the characteristics then shows

$$u_1(x, t) = \cosh(x - t) + \sinh(x + t),$$
$$u_2(x, t) = \cosh(x - t) - \sinh(x + t).$$

It is a simple matter to verify that these results both satisfy the original equations (81) and the initial conditions (82) so that they are, indeed, the solution to our problem. See §109, question 17, for the connection that exists between this method of solution and d'Alembert's method of solution for the wave equation.

§106   Simple waves

When one of the Riemann invariants r or s is *identically constant*, the corresponding solutions of equations (57) of §105 are known as **simple wave solutions**. That is, simple wave solutions occur either when $r(\beta) \equiv r_0 = $ const., or when $s(\alpha) \equiv s_0 = $ const., and we now deduce the basic properties of this fundamental class of solutions directly from this simple definition.

Suppose, for example, that $s(\alpha) \equiv s_0$, then equations (67) and (68) may be written

$$f_{11}(u_1) + f_{12}(u_2) = r(\beta) \quad \text{along } C^{(1)} \text{ characteristics} \tag{84}$$

and

$$f_{21}(u_1) + f_{22}(u_2) = s_0 \quad \text{along } C^{(2)} \text{ characteristics,} \tag{85}$$

where

$$f_{ij}(u_j) = \int \mu l_j^{(i)} \, du_j. \tag{86}$$

This shows that everywhere along a $C^{(1)}$ characteristic specified by $\beta = \beta_0 = $ const., say, u_1 and u_2 must also be constant, for they are the solution of the nonlinear system of simultaneous equations

$$f_{11}(u_1) + f_{12}(u_2) = r(\beta_0),$$

and

$$f_{21}(u_1) + f_{22}(u_2) = s_0.$$

The actual constant values associated with u_1 and u_2 along this characteristic are $u_1 = \tilde{u}_1(\xi)$, $u_2 = \tilde{u}_2(\xi)$ determined by the values of the initial date (58) at the point $(\xi, 0)$ of the initial line through which this $C^{(1)}$ characteristic passes. Any function of u_1 and u_2 will also be constant along this characteristic as, in particular, will be $\lambda^{(1)}(\tilde{u}_1(\xi), \tilde{u}_2(\xi)) = \Lambda^{(1)}(\xi)$, say. Consequently, as the $C^{(1)}$ characteristic is found from (69) by integrating

$$C^{(1)}: \frac{dx}{dt} = \Lambda^{(1)}(\xi),$$

it must be the straight line

$$x = \xi + t\Lambda^{(1)}(\xi). \tag{87}$$

As β_0 and hence ξ, were arbitrary, this result implies that by allowing ξ to move along its permitted interval on the initial line, so (87) will generate a *straight line family of $C^{(1)}$ characteristics*. Conversely, had we set $r(\beta) \equiv r_0$, it would then have followed that the $C^{(2)}$ family of characteristics was a family of straight lines along each of which u_1 and u_2 were constant.

If the $C^{(1)}$ family of characteristics converges they will generate an envelope in the (x, t)-plane at each point of which the solution u_1 and u_2 will become non-unique, as in §102. This depends on the function $\Lambda^{(1)}(\xi)$, and the envelope, when it occurs for $t > 0$, is then given as in §102 by solving the pair of equations

$$\xi + t\Lambda^{(1)}(\xi) - x = 0,$$

and

$$1 + t\frac{d\Lambda^{(1)}}{d\xi} = 0. \tag{88}$$

When this envelope is required, rather than attempting to eliminate ξ between these two equations, it is usually simpler to solve for x and t in terms of ξ as a parameter. A corresponding situation exists when $r(\beta) \equiv r_0$ and it is the $C^{(2)}$ family of characteristics that comprise the straight line family.

By analogy with the situation in gas dynamics, when a straight line family of characteristics converges, the associated simple wave is often called a **compression wave**, whereas when it diverges, the associated simple wave is called an **expansion wave**.

The property that in a simple wave u_1 and u_2 are constant along the straight line characteristics means that simple wave solutions are the ones that must occur adjacent to a region of constant state, as first discussed in §104 (see also §109, question 19). This fundamental property of simple waves makes them useful when piecing together solutions to more complicated problems, as will be illustrated in the next section.

§107   The piston problem

The actual use of a simple wave solution to piece together a more complicated solution is well illustrated by the so-called *piston problem* in gas dynamics. This one-dimensional unsteady problem involves determining the gas motion induced in a semi-infinite tube filled with gas, that is initially at rest, when a piston closing one end is caused to move. If the piston is withdrawn from the tube in a smoothly accelerated manner for a time t_1, after which the speed of withdrawal remains constant, the piston path will follow a curve in the (x, t) plane like the dotted line in Fig. 39(a).

In Fig. 39(a) the initial region in which the gas is at rest, so that there $u = 0$, $\rho = \rho_0$, $a = a_0$, is denoted by (I), and the characteristic $C_0^{(1)}$ that bounds it and passes through the origin is obtained by integrating

$$\frac{dx}{dt} = (u + a)_0 = a_0,$$

so that $C_0^{(1)}$ has the equation $x = a_0 t$.

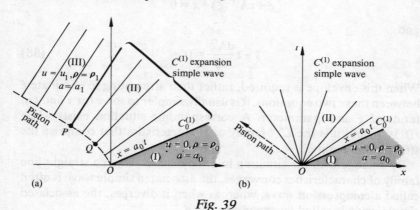

Fig. 39

If point P is the point on the piston path after which the piston withdrawal velocity remains constant at V, say, then region (II) is a *simple wave region* and it is traversed by a family of straight line $C^{(1)}$

characteristics, each of which issues out from a point of the piston path between O and P. The gradient of a typical member of these characteristics through point Q will be the sum of the sound speed a and the piston velocity at Q, since unless there is a vacuum at the piston face the piston and gas velocity must be the same. As the piston velocity remains constant after P, the straight line $C^{(1)}$ characteristics traversing region (III) will all be parallel, indicating that this is a *constant state region* in which $u = u_1$, $\rho = \rho_1$, and $a = a_1$. This has thus provided a physical example of the way in which a simple wave region connects two different constant state regions (I) and (III).

A different situation arises if the piston is suddenly withdrawn with the constant velocity V. The piston path in this case is the straight dotted line with gradient V that is shown in Fig. 39(b). In this case the family of straight line $C^{(1)}$ characteristics must all radiate out from the single point O in order to traverse the simple wave region (II). Such simple waves are called **centred simple waves**. The bounding characteristic $C_0^{(1)}$ through O will, of course, again be the straight line $x = a_0 t$.

If the piston starts from rest and is pushed into the gas in a smoothly accelerated manner, as shown in Fig. 40, the $C^{(1)}$ characteristics

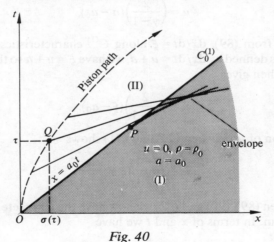

Fig. 40

originating from points on the piston path will converge in region (II) and form an envelope starting at P. The shape of this envelope will be described by equations (88) in **§106**. Point P represents the start of a *gas shock*, across which gas velocity and density are discontinuous.

Region (I) bounded by the characteristic $C_0^{(1)}$ with the equation $x = a_0 t$ will be as before. We return to this problem at the end of this section when we will determine the precise location of P on the characteristic $C_0^{(1)}$.

Let us now examine in detail the centred simple wave shown in Fig. 39(b). The Riemann invariants for this case have already been found in equations (77) and (78) of §105. The family of straight line $C^{(1)}$ characteristics through O that traverse region (II) have the equation

$$x/t = \xi, \tag{89}$$

where ξ is a parameter, with $C_0^{(1)}$ corresponding to $\xi = a_0$.

All the $C^{(2)}$ characteristics must enter the constant state region (I) in which $u = 0$, $\rho = \rho_0$, $a = a_0$, so that the Riemann invariant $s(\alpha)$ in (78) must be identically constant and of the form

$$\frac{2a}{\gamma - 1} - u = \frac{2a_0}{\gamma - 1}.$$

This shows that

$$u = \left(\frac{2}{\gamma - 1}\right)(a - a_0), \tag{90}$$

but since, from (89), $dx/dt = \xi$ along $C^{(1)}$ characteristics, which are themselves defined by $dx/dt = u + a$, we have $\xi = u + a$ so that elimination of a then gives

$$u = \left(\frac{2}{\gamma + 1}\right)(\xi - a_0). \tag{91}$$

Elimination of u between (90) and (91) shows

$$a = \left(\frac{\gamma - 1}{\gamma + 1}\right)\xi + \frac{2a_0}{\gamma + 1}, \tag{92}$$

and between (89), (90) and (92) we now have the complete solution to our problem. In terms of x and t we have

$$u = \left(\frac{2}{\gamma + 1}\right)\left(\frac{x}{t} - a_0\right),$$

and

$$a = \left(\frac{\gamma - 1}{\gamma + 1}\right)\left(\frac{x}{t}\right) + \frac{2a_0}{\gamma + 1},$$

which specifies u and a, and hence ρ, at points in simple wave region (II) of Fig. 39(b).

As $a^2 = dp/d\rho = \gamma k\rho^{\gamma-1}$, it follows that $\rho = 0$ when $a = 0$, so that setting $a = 0$ on (92) shows that the critical value $\xi = \xi_c$ at which this happens is

$$\xi_c = -\frac{2a_0}{\gamma - 1}.$$

The corresponding critical speed u_c at which the gas density reduces to zero is, then, from (91),

$$u_c = -\frac{2a_0}{\gamma - 1}.$$

This speed is called the **cavitation speed** and if the piston withdrawal speed V exceeds u_c there will be a third region (III) in the flow in which there is a *vacuum*. Such a situation is shown in Fig. 41 in which from (89) and (96) the cavitation line has the equation $x = -2a_0t/(\gamma-1)$. It

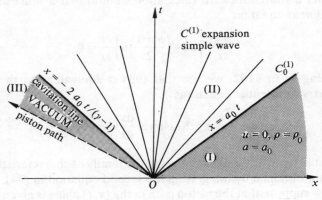

Fig. 41

should be remarked that the non-uniqueness of the solution at the origin in this centred simple wave is immediately resolved in this expansion process. The term *expansion* that is used here derives from the fact that as ξ increases from $\xi = a_0$, corresponding to $C_0^{(1)}$, so the gas velocity u increases and the sound speed a, and hence the density ρ, decreases. The converse occurs in the situation illustrated in Fig. 40, so that there the physical process corresponds to a *compression wave*.

To complete our examination of the piston problem let us now determine the location of the point P, with coordinates (x_P, t_P), at which a shock first forms on the advancing wavefront $C_0^{(1)}$ when the piston is pushed into the gas.

It follows directly from (91), and the fact that $a^2 = \gamma k \rho^{\gamma-1}$, that throughout the simple wave region (II) in Fig. 40,

$$a = a_0 + \left(\frac{\gamma-1}{2}\right)u,$$

and

$$\rho = \rho_0 \left\{ 1 + \left(\frac{\gamma-1}{2a_0}\right)u \right\}^{2/(\gamma-1)}$$

These equations relate both a and ρ throughout the simple wave region (II) to the single variable u. To find out how u varies let us now substitute a and ρ into the equation of conservation of mass given in **§100**, equation (2) (equation (3) would serve equally well).

After a straightforward calculation we find that u must satisfy the quasilinear equation

$$\frac{\partial u}{\partial t} + \left\{ a_0 + \left(\frac{\gamma+1}{2}\right)u \right\}\frac{\partial u}{\partial x} = 0. \tag{93}$$

Arguing as in **§102**, we see that (93) is equivalent to the pair of ordinary differential equations

$$\frac{\mathrm{d}u}{\mathrm{d}t} = 0 \quad \text{along the curves } \frac{\mathrm{d}x}{\mathrm{d}t} = a_0 + \left(\frac{\gamma+1}{2}\right)u, \tag{94}$$

so that $u = \text{const.}$, along members of the family of characteristic curves that are obtained by integrating the second equation in (94).

Now suppose that the piston path in the (x, t) plane is given in terms of time by the equation $x = \sigma(t)$, with $\sigma(0) = 0$ and $(\mathrm{d}\sigma/\mathrm{d}\tau)_{\tau=0} = \sigma'(0) = 0$, so that the piston starts from rest at the origin. Then at the point Q on the piston path in Fig. 40 with coordinates $(\sigma(\tau), \tau)$, the piston velocity will be $\sigma'(\tau)$. As $u = \text{const.}$, along the characteristics of equation (93), and u equals the piston velocity at Q, the equation of the straight line characteristic through Q obtained by integrating the second equation of (94) is

$$x = \sigma(\tau) + \left\{ a_0 + \left(\frac{\gamma+1}{2}\right)\sigma'(\tau) \right\}(t-\tau). \tag{95}$$

As in §102, the envelope of this family of straight line characteristics, with τ as parameter, is obtained by eliminating τ between (95) and the equation that is obtained when it is differentiated partially with respect to τ to yield

$$0 = \sigma'(\tau) + (t - \tau)\left(\frac{\gamma + 1}{2}\right)\sigma''(\tau) - \left\{a_0 + \left(\frac{\gamma + 1}{2}\right)\sigma'(\tau)\right\}. \quad (96)$$

Since our concern is only with the location of the start of the envelope at point P on the characteristic $C_0^{(1)}$ through the origin, and not with the entire envelope, we require to take point Q to be at the origin. This is equivalent to setting $\tau = 0$, when from (95), (96) and the initial conditions for $\sigma(\tau)$ and $\sigma'(\tau)$ we discover that

$$t_P = \frac{2a_0}{(\gamma + 1)\sigma''(0)} \quad \text{and} \quad x_P = a_0 t_P. \quad (97)$$

This is the required result and shows that the time of shock formation t_P is determined by the initial acceleration $\sigma''(0)$ of the piston. If $\sigma''(0) > 0$ the shock will form a finite time after the start of the motion, but if $\sigma''(0) < 0$ then no shock will form. The former case corresponds to pushing the piston into the gas, and the latter to withdrawing the piston. If the piston is pushed in impulsively, so that the initial acceleration is infinite, then (97) shows that the shock will form immediately on the piston face.

§108   Discontinuous solutions and shock waves

The development of a non-unique solution to a quasilinear hyperbolic equation has already been encountered in several different contexts. In the physical world this non-uniqueness usually manifests itself in the formation of solutions which are *discontinuous* across some surface. In gas dynamics these discontinuous solutions are called **shocks**, and across a shock the gas velocity and density change discontinuously. When discontinuities of this type propagate in a gas they are called **shock waves**. By analogy, the term shock wave is also often used even when the discontinuities occur in media other than gases as, for example, when discontinuous stress waves propagate in solids. In the remainder of this chapter we shall examine in fairly general terms the type of discontinuous solution that is allowed by systems of conservation laws.

The starting point will be the general scalar conservation equation in three space dimensions and time which may be written

$$\frac{\partial u}{\partial t} + \text{div } \mathbf{f} = 0, \tag{98}$$

where the scalar $u = u(\mathbf{r}, t)$ is a function of the position vector $\mathbf{r}$ and time t, and $\mathbf{f} = \mathbf{f}(u)$ is a vector function of u. To deduce the nature of the *discontinuous* solutions u that are permissible in (98) we first establish a general theorem. This will be required in order to allow for the possibility that the discontinuity, or shock, may occur across a moving surface.

Consider an arbitrary surface $S(t)$, moving with velocity $\mathbf{q}$, that bounds a volume $V(t)$ in which a differentiable scalar function u is defined.

Then setting

$$I = \int_{V(t)} u \, dV, \tag{99}$$

we notice that I will change in the time increment δt both because u is time dependent and because the volume $V(t)$ bounded by $S(t)$ will change as $S(t)$ moves. To the first order, in the time increment δt the integrand of I changes from u to

$$u + \left(\frac{\partial u}{\partial t}\right) \delta t.$$

To deduce the effect of the change of volume we also notice that the vector surface element $d\mathbf{S}$ of $S(t)$ moves a distance $\mathbf{q} \, \delta t$ in time increment δt, so that the corresponding element of volume change due to movement of the surface must be $\mathbf{q} \cdot d\mathbf{S} \, \delta t$. The increment in I due to this change will be $u\mathbf{q} \cdot d\mathbf{S} \, \delta t$. Adding these results and integrating the two separate contributions over $V(t)$ and $S(t)$, respectively, gives

$$I + \delta I = \int_{V(t)} \left\{ u + \left(\frac{\partial u}{\partial t}\right) \delta t \right\} dV + \int_{S(t)} u\mathbf{q} \cdot d\mathbf{S} \, \delta t.$$

Substracting equation (99) from this result, dividing by δt and proceeding to the limit as $\delta t \to 0$, finally leads to the rate of change theorem

$$\frac{D}{Dt} \int_{V(t)} u \, dV = \int_{V(t)} \left(\frac{\partial u}{\partial t}\right) dV + \int_{S(t)} u\mathbf{q} \cdot d\mathbf{S}. \tag{100}$$

Here the notation D/Dt has been used to denote the total derivative as seen by an observer moving with velocity $\mathbf{q}$ (see §46, Chapter 5).

Application of the Gauss divergence theorem to result (100) enables it to be expressed in the following alternative form which is more convenient for our purposes

$$\frac{D}{Dt} \int_{V(t)} u \, dV = \int_{V(t)} \left(\frac{\partial u}{\partial t} + \text{div} \, (u\mathbf{q}) \right) dV. \tag{101}$$

We now use this theorem to derive the *discontinuity conditions*, or the so-called *jump conditions*, for a solution u to the scalar conservation law (98). In doing so we assume that a discontinuity surface exists and that an arbitrary part of it, $S^*(t)$, divides the volume $V(t)$ into volumes $V_0(t)$ and $V_1(t)$, and the boundary surface $S(t)$ of $V(t)$ into the two surfaces $S_0(t)$ and $S_1(t)$, respectively. The value of functions on adjacent sides of and arbitrarily close to $S^*(t)$ will be denoted by the suffixes 0 and 1, according as $S^*(t)$ is approached from $V_0(t)$ or $V_1(t)$, respectively.

Using the expression for $\partial u/\partial t$ from (98) in theorem (101), applying the Gauss divergence theorem again and assuming that u has no singularities other than the discontinuity then gives

$$\frac{D}{Dt} \int_{V(t)} u \, dV = \int_{S(t)} (u\mathbf{q} - \mathbf{f}) \cdot d\mathbf{S}. \tag{102}$$

Substracting from this equation the corresponding equations in which $V(t)$ is identified, respectively, with $V_0(t)$ and $V_1(t)$, and the surface $S(t)$ with the corresponding surfaces bounding these volumes, shows that

$$\int_{S^*(t)} (u\mathbf{q} - \mathbf{f})_0 \cdot d\mathbf{S}_0^* + \int_{S^*(t)} (u\mathbf{q} - \mathbf{f})_1 \cdot d\mathbf{S}_1^* = 0, \tag{103}$$

where $d\mathbf{S}_i^*$ is the outwardly directed vector surface element of $S^*(t)$ with respect to $V_i(t)$. However,

$$d\mathbf{S}_0^* = -d\mathbf{S}_1^* = \mathbf{n} \, dS^*,$$

say, where $\mathbf{n}$ is the outward drawn unit normal with respect to $V_0(t)$. Thus as $S^*(t)$ is an arbitrary part of the discontinuity surface it follows directly from (103) that across $S^*(t)$

$$(u\mathbf{q} - \mathbf{f}(u))_0 \cdot \mathbf{n} - (u\mathbf{q} - \mathbf{f}(u))_1 \cdot \mathbf{n} = 0. \tag{104}$$

Employing the notation $[\![\alpha]\!]$ to denote the discontinuous jump $\alpha_0 - \alpha_1$ in the quantity α across any point of $S^*(t)$ allows (104) to be re-expressed as

$$[\![\tilde{\lambda} u - \mathbf{n} \cdot \mathbf{f}(u)]\!] = 0, \tag{105}$$

in which

$$\tilde{\lambda} = \mathbf{n} \cdot \mathbf{q} \tag{106}$$

is the *speed of propagation of the discontinuity along the normal* to $S^*(t)$.

This is the general result we were seeking in respect of discontinuous, or shock, solutions to conservation law (98). It is an algebraic relationship that must be satisfied at each point of the discontinuity surface and it connects the values of u on adjacent sides of the surface with the speed of propagation $\tilde{\lambda}$ of an element of the surface along its normal. It should, however, be observed that although (105) describes the magnitude of the jump $[\![u]\!]$, because of the nonlinearity it does not necessarily determine the sense of the jump across $S^*(t)$.

If instead of a scalar conservation law like (98) a system of simultaneous conservation laws is involved, result (105) must be applied individually to each one. The resulting system of algebraic jump conditions must then hold simultaneously. It is important at this stage to recognise that the discontinuity or shock speeds $\tilde{\lambda}$ that are permitted in such a system will *not*, in general, be the same as the characteristic speeds λ, though for a system of n conservation laws they will both equal n in number. Only when the conservation laws are linear constant coefficient equations will the $\tilde{\lambda}$ and the λ always coincide.

To see this consider the one-dimensional system of conservation laws

$$\frac{\partial \mathbf{U}}{\partial t} + \frac{\partial \mathbf{F}}{\partial x} = \mathbf{0}, \tag{107}$$

as in (14), but with $\mathbf{F} = \mathbf{A}\mathbf{U}$ and with $\mathbf{A}$ an $n \times n$ constant coefficient matrix. Then system (107) is equivalent to

$$\frac{\partial \mathbf{U}}{\partial t} + \mathbf{A} \frac{\partial \mathbf{U}}{\partial x} = \mathbf{0},$$

and so will be hyperbolic if the n eigenvalues $\lambda^{(i)}$ satisfying

$$|\mathbf{A} - \lambda \mathbf{I}| = 0 \tag{108}$$

are all real and the corresponding n eigenvectors (left or right) are linearly independent.

Now in one space dimension the discontinuity surface, or shock, will be planar and (105) reduces to

$$[\![\tilde{\lambda}u - f(u)]\!] = 0,\tag{109}$$

so that when this condition is applied simultaneously to each of the n conservation laws in system (107) it is easily seen that we arrive at a system of equations that can be put into the form

$$(\mathbf{A} - \tilde{\lambda}\mathbf{I})[\![\mathbf{U}]\!] = \mathbf{0}.\tag{110}$$

For this homogeneous system to have a non-trivial discontinuous solution we must require that

$$|\mathbf{A} - \tilde{\lambda}\mathbf{I}| = 0.$$

This shows, by comparison with (108), that in this case $\lambda \equiv \tilde{\lambda}$, and that both are the eigenvalues of the constant coefficient matrix $\mathbf{A}$. The jump vectors $[\![\mathbf{U}]\!]$ are then proportional to the corresponding right eigenvectors of $\mathbf{A}$.

Returning once more to systems of n quasilinear conservation laws, but now considering only the one-dimensional case, we see from (105) that discontinuous or shock solutions to the conservation system

$$\frac{\partial \mathbf{U}}{\partial t} + \frac{\partial \mathbf{F}}{\partial x} = \mathbf{0}\tag{111}$$

must satisfy the matrix jump conditions

$$[\![\tilde{\lambda}\mathbf{U} - \mathbf{F}(\mathbf{U})]\!] = \mathbf{0},\tag{112}$$

where, as previously, $\mathbf{U}$ and $\mathbf{F}$ are n element column vectors. It follows from (106) that $\tilde{\lambda}$ is the speed of propagation of the planar discontinuity surface, or shock, along its normal.

Because of a special result in gas dynamics which we discuss later, the general result (112) is often called the **generalised Rankine–Hugoniot relation** that is to be satisfied by a discontinuous solution to system (111).

To conclude this section let us now apply these results to one-dimensional gas shocks. The equations that govern the gas flow are:

conservation of mass

$$\frac{\partial \rho}{\partial t} + \frac{\partial (\rho u)}{\partial x} = 0,\tag{113}$$

conservation of momentum

$$\frac{\partial(\rho u)}{\partial t} + \frac{\partial}{\partial x}(\rho u^2 + p) = 0, \tag{114}$$

conservation of energy

$$\frac{\partial}{\partial t}(\tfrac{1}{2}\rho u^2 + \rho e) + \frac{\partial}{\partial x}\{u(\tfrac{1}{2}\rho u^2 + \rho e + p)\} = 0. \tag{115}$$

Equations (113) and (114) have both been encountered before, but equation (115) is new and describes the energy of the gas in terms of its density ρ, the gas velocity u, the gas pressure p and the specific internal energy e of the gas. The form of e depends on the nature of the gas, but for most purposes it is justifiable to assume a perfect gas and to set

$$e = \frac{p}{\rho(\gamma - 1)}, \tag{116}$$

where γ is the adiabatic exponent for the gas.

Applying (105) to each of these results then shows that the jump conditions across a gas shock are determined by

conservation of mass

$$[\![\tilde{\lambda}\rho - \rho u]\!] = 0, \tag{117}$$

conservation of momentum

$$[\![\tilde{\lambda}\rho u - (\rho u^2 + p)]\!] = 0, \tag{118}$$

conservation of energy

$$[\![\tilde{\lambda}(\tfrac{1}{2}\rho u^2 + \rho e) - u(\tfrac{1}{2}\rho u^2 + \rho e + p)]\!] = 0. \tag{119}$$

It proves convenient to re-express these results in terms of

$$\tilde{u} = u - \tilde{\lambda}, \tag{120}$$

which is the gas speed relative to an observer moving with the shock at the speed $\tilde{\lambda}$. These results the become

conservation of mass

$$[\![\rho\tilde{u}]\!] = 0, \tag{121}$$

conservation of momentum

$$[\![\rho\tilde{u}u + p]\!], \tag{122}$$

conservation of energy

$$[\![\tilde{u}(\tfrac{1}{2}\rho u^2 + \rho e) + up]\!] = 0. \tag{123}$$

In this form (121) may be interpreted as asserting that the mass flow m through a unit area of the shock in a unit time is constant, so that

$$m = \rho_0\tilde{u}_0 = \rho_1\tilde{u}_1, \tag{124}$$

where the suffixes 0, 1 are used to denote quantities on adjacent sides of the discontinuity surface. In gas dynamics a discontinuity surface is only called a shock when there is a mass flow across it, so that fluid particles actually cross a gas shock. It is conventional to refer to the side of a gas shock through which gas enters as the **front** of the shock or as the side **ahead** of the shock. The other side is called the **back** of the shock or the side **behind** the shock. We shall take the suffix 0 to refer to the side ahead of the shock and the suffix 1 to refer to the side behind the shock.

We remark here that discontinuity surfaces across which no mass flow takes place can also occur in fluid mechanics. These are called **contact discontinuities**, and they separate fluids belonging to different thermodynamic states in which the flow is tangential to the discontinuity surface.

The jump conditions (121) to (123) have been derived quite generally for an element of a plane shock wave moving with the local speed $\tilde{\lambda}$ along its normal. In a steady state situation the value attributed to $\tilde{\lambda}$ will determine how the shock moves relative to the reference frame in which $\tilde{\lambda}$ is measured. If, for example, we set $\tilde{\lambda} = 0$ the shock will be stationary, whereas if we set $\tilde{\lambda} = -\tilde{u}_0$, then $u_0 = 0$ and the shock will propagate with speed $\tilde{u}_0$ into the gas of region 0 which is then at rest.

When the gas is perfect, so that the specific internal energy e is determined by expression (116), use of (124) enables us to re-write the energy equation (123) in the form

$$\tfrac{1}{2}m[\![u^2]\!] + \frac{m}{\gamma - 1}[\![p\tau]\!] + [\![up]\!] = 0, \tag{125}$$

where $\tau = 1/\rho$ is called the specific volume of the gas. If, now, we multiply the momentum equation (122) by $\tfrac{1}{2}(u_0 + u_1)$, which is the average of the gas speeds on either side of the shock, and again use (124) we find

$$\tfrac{1}{2}m[\![u^2]\!] = -\tfrac{1}{2}(u_0 + u_1)[\![p]\!],$$ (126)

so that (125) becomes

$$\frac{m}{\gamma - 1}[\![p\tau]\!] - \tfrac{1}{2}(u_0 + u_1)[\![p]\!] + [\![up]\!] = 0.$$ (127)

As the speed $\tilde{\lambda}$ of the shock must be continuous across the shock it follows from (119) that

$$[\![\tilde{u}]\!] = [\![u]\!]$$ (128)

or, equivalently, that

$$m[\![\tau]\!] = [\![u]\!].$$ (129)

So, applying the identity

$$[\![PQ]\!] = \tfrac{1}{2}(P_0 + P_1)[\![Q]\!] + \tfrac{1}{2}(Q_0 + Q_1)[\![P]\!]$$

to the second and third terms of (127), using (129) and cancelling the non-zero factor m, finally reduces it to

$$\frac{1}{\gamma - 1}[\![p\tau]\!] + \tfrac{1}{2}(p_0 + p_1)[\![\tau]\!] = 0.$$ (130)

We remark, in passing, that this relationship, which is equivalent to the more general result

$$[\![e + \tfrac{1}{2}(p_0 + p_1)\tau]\!] = 0,$$ (131)

is the thermodynamic relationship known in gas dynamics as the **Rankine–Hugoniot relation**. It is, of course, for this reason that the name generalised Rankine–Hugoniot relation was given to the general matrix jump condition (112).

By introducing the ratio $r = \tau_0/\tau_1 (= \rho_1/\rho_0)$, where as before the suffixes $0, 1$ refer to conditions ahead of and behind the shock, respectively, it is a simple matter to show from (130) that

$$\frac{p_1}{p_0} = \frac{(\gamma + 1)r - (\gamma - 1)}{(\gamma + 1) - (\gamma - 1)r}.$$ (132)

This result is important because it relates the four quantities p_0, p_1, ρ_0, ρ_1 across a shock and enables any one of them to be determined in terms of the other three. As the pressure ratio p_1/p_0 is inherently positive, the numerator and denominator of (132) must be of the same sign, so that as $\gamma > 1$ we arrive at the condition

$$\frac{\gamma - 1}{\gamma + 1} < r < \frac{\gamma + 1}{\gamma - 1}. \tag{133}$$

Inspection of this inequality reveals that from the mathematical point of view the density ratio r across a shock may assume values both less and greater than unity. When $r > 1$ the shock will then involve a **compression**, but when $r < 1$ it will involve an **expansion**, or **rarefaction**. Clearly in any physical situation the sense of change of the density and the corresponding pressure jump across a shock will be uniquely determined, yet as it stands (133) will allow either a compression or a rarefaction shock to occur. This ambiguity involving which of the two types of mathematically possible shock should be used must be resolved if our result is to be related to the physical world.

Expressed differently, we have found that in a given situation it is mathematically possible for either a compression or a rarefaction gas shock to occur as a solution to the jump conditions, but the uniqueness we expect of situations in the physical world demands that somehow we choose between them. As rarefaction shocks have not been observed experimentally we must reject them as non-physical and, accordingly, confine r to the interval.

$$1 \leqslant r < \frac{\gamma + 1}{\gamma - 1}. \tag{134}$$

This condition, which has been proposed here on the basis of experimental evidence, is now sufficient to ensure uniqueness for gas shock solutions to the jump conditions (121) to (123).

It is, indeed, a feature of the generalised Rankine-Hugoniot relation (112) that to obtain a unique shock solution it is necessary to supplement the relation by some further condition such as (134). Although it will not be done here, it is not difficult to show that in gas dynamics the physically meaningful compression shock can be selected simply by appeal to the thermodynamical requirement that *the entropy of the gas flow should not decrease across a shock*. That is, this entropy inequality is equivalent to condition (134), since across a rarefaction shock the entropy would decrease, thereby violating the second law of thermodynamics.

In other physical situations different criteria must be used to select the physically meaningful shock solutions from the mathematically possible ones. Usually these amount either to an energy inequality or to the assumption that only a physically meaningful shock is stable with respect to small disturbances. For example, when studying the behaviour of a bore in a river (see §102), the inequality condition that is used in place of (134) is that since there is no energy source in the discontinuity surface that represents the bore, fluid cannot gain energy when passing through such a discontinuity surface. An idealised version of this situation is shown in Fig. 42 where the bore is moving to the left from water of depth h_1 into water of depth h_0 (see §109 question (23)).

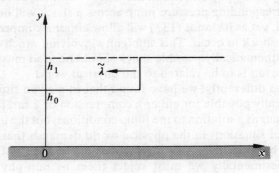

Fig. 42

§109   Examples

1. Give an alternative to the new variables u, v introduced in §100 to reduce equation (1) to a quasilinear system of two first order equations.

2. Use matrix notation to write out the system of equations (2) and (3) when u is a three dimensional vector with components u, v, w. Define all matrices that are used.

3. Use matrix notation to write out the one-dimensional form of Maxwell's equations

$$\operatorname{curl} \mathbf{E} = -\mu_0 \varkappa_m \frac{\partial \mathbf{H}}{\partial t} \quad \text{and} \quad \operatorname{curl} \mathbf{H} = \varepsilon_0 \varkappa_e \frac{\partial \mathbf{E}}{\partial t}$$

when $\mathbf{E}$ and $\mathbf{H}$ depend only on the x coordinate and the time t.

4. Express the result of the above question as a matrix conservation law

$$\frac{\partial \mathbf{U}}{\partial t} + \frac{\partial \mathbf{F}}{\partial x} = \mathbf{0},$$

defining the four element column vectors $\mathbf{U}$ and $\mathbf{F}(\mathbf{U})$ that are involved.

5. The one-dimensional long wave approximation for water waves leads to the equations

$$\frac{\partial u}{\partial t} + u\frac{\partial u}{\partial x} + 2c\frac{\partial c}{\partial x} - \frac{dH}{dx} = 0,$$

$$2\frac{\partial c}{\partial t} + 2u\frac{\partial c}{\partial x} + c\frac{\partial u}{\partial x} = 0,$$

Here u is the horizontal velocity component, $c = \sqrt{(gh)}$ is the surface wave propagation speed, with h the depth of the water and g the acceleration due to gravity, and $H(x) = gY(x)$, with $y + Y(x) = 0$ the equation of the sea-bed relative to an origin in the equilibrium surface. By inspection in the case of the first equation, and by inspection after multiplication of the second equation by a simple factor, express the equations as a matrix conservation law.

6. Derive the parametric equations of the envelope given in §102, equation (27).

7. Find the parametric form of the characteristic curves and the implicit solution of

$$\frac{\partial u}{\partial t} + u\frac{\partial u}{\partial x} = 0$$

when $u(x, 0) = \sin x$. Hence find and sketch the envelope of the characteristics for $t > 0$ in the interval $0 < x \leqslant 2\pi$. Show that this equation can be written as a conservation law. Compare this conservation law with §101, equation (15), when ρ and p are constant.

8. Consider the equation

$$\frac{\partial u}{\partial t} + \tanh u\frac{\partial u}{\partial x} = 0,$$

when

(a) $u(x, 0) = \tanh^{-1} x$

(b) $u(x, 0) = -\tanh^{-1} x$.

Show that in case (a) the solution is defined for all x and $t > 0$, but that in case (b) the solution is not defined for $t \geq 1$.

9. Show that the system of equations (7) and (8) in §**100** is totally hyperbolic and that the wave propagation speeds are

$$\lambda = \pm c / (1 + (\partial y / \partial x)^2)$$

10. By writing equation (6) of §**64** in the form of a first order quasilinear system, show that the equation governing sound waves in a gas is totally hyperbolic. Hence show that the speeds of wave propagation described by this equation are

$$\lambda = \pm \sqrt{(\gamma p_0 / \rho_0)} [1 + \partial \xi / \partial x]^{-(\gamma+1)/2}$$

11. Write the equations in question (5) above in the form of a system and hence show that the long wave approximation for water waves is totally hyperbolic. Show also that the speeds of wave propagation are

$$\lambda = u \pm c.$$

12. Steady two-dimensional irrotational isentropic gas flow is governed by the equations

$$\frac{\partial v}{\partial x} - \frac{\partial u}{\partial y} = 0,$$

$$(a^2 - u^2)\frac{\partial u}{\partial x} - uv\left(\frac{\partial u}{\partial y} + \frac{\partial v}{\partial x}\right) + (a^2 - v^2)\frac{\partial v}{\partial y} = 0,$$

where (u, v) are the components of the fluid velocity $\mathbf{q}$ in the x and y directions, and a is the local speed of sound. Show that these equations are only totally hyperbolic in the (x, y) plane when the flow $\mathbf{q}$ is supersonic, so that

$$u^2 + v^2 > a^2.$$

By defining the *Mach number* $M = (u^2 + v^2)^{1/2}/a$, show that the two characteristics passing through a point in a supersonic flow are both inclined to the fluid velocity vector $\mathbf{q}$ at an angle α, where $\sin \alpha = 1/M$.

13. The equations of unsteady one-dimensional non-isentropic gas flow are

$$\frac{\partial \rho}{\partial t} + u\frac{\partial \rho}{\partial x} + \rho\frac{\partial u}{\partial x} = 0,$$

$$\frac{\partial u}{\partial t} + u\frac{\partial u}{\partial x} + \frac{a^2}{\rho}\frac{\partial \rho}{\partial x} + \frac{1}{\rho}\frac{\partial p}{\partial S}\frac{\partial S}{\partial x} = 0,$$

$$\frac{\partial S}{\partial t} + u\frac{\partial S}{\partial x} = 0,$$

where u is the gas velocity in the x direction, ρ is the density, S is the entropy, $p = p(\rho, S)$ is the pressure and $a^2 = \partial p/\partial \rho$ is the square of the local sound speed. Show that the system is totally hyperbolic and that the three speeds of wave propagation are

$$\lambda^{(1)} = u + a, \qquad \lambda^{(2)} = u - a, \qquad \lambda^{(3)} = u.$$

Interpret physically the nature of the wave propagation associated with $\lambda = \lambda^{(3)}$.

14. Consider question (5) above in which $H = $ const. Show that in the disturbed region immediately adjacent to a wavefront bounding a constant state, either $dc = \frac{1}{2}\,du$, or $dc = -\frac{1}{2}\,du$. Interpret the meaning of these wavefronts in physical terms.

15. Modify the rule derived in §104 so that it applies to steady state problems in the (x, y) plane, and then apply it to the equations in question (12) above for a supersonic flow adjacent to a constant state region. Interpret your results in physical terms.

16. Complete the details of the calculations involved in the solution of equations (81) of §105 when subject to the initial data (82).

17. By differentiation and elimination show that u_1 and u_2 in equations (81) of §105 are each solutions of the wave equation

$$\frac{\partial^2 \phi}{\partial t^2} = \frac{\partial^2 \phi}{\partial x^2}.$$

Use equations (81) and the initial data (82) to derive appropriate initial data for u_1 and u_2 as solutions of the wave equation and hence find u_1 and u_2 by means of d'Alembert's method as described in §11 equation (57). Compare and contrast this method of solution with the one described in §105 making use of Riemann invariants.

18. Use the method of Riemann invariants to show that the solution to equations (81) of §105 subject to the initial data

$$u_1(x, 0) = 1, \qquad u_2(x, 0) = \sin x$$

is

$$u_1(x, t) = 1 + \tfrac{1}{2}[\sin (x - t) - \sin (x + t)],$$
$$u_2(x, t) = \tfrac{1}{2}[\sin (x - t) + \sin (x + t)].$$

19. Show, using the equations of question (5) above, that when the sea-bed is horizontal the Riemann invariants for the long water wave approximation are

$$2u + c = r(\beta) \quad \text{along } C^{(1)} \text{characteristics,}$$

and

$$2u - c = s(\alpha) \quad \text{along } C^{(2)} \text{characteristics.}$$

20. Show by working directly with the Riemann invariants for one-dimensional unsteady isentropic gas flow, as described by equations (10), (11) of §100, that adjacent to a constant state (u_0, a_0, ρ_0) bounded by a $C^{(1)}$ characteristic, $du = (a_0/\rho_0) \, d\rho$. Show also that if this state is bounded by a $C^{(2)}$ characteristic then $du = -(a_0/\rho_0) \, d\rho$. Hence verify the conclusions of §104.

21. Show that the path followed by a gas particle in the centred simple wave flow discussed at the end of §107 has the equation

$$x = \left(\frac{a_0 t}{\gamma - 1}\right)\left[(\gamma + 1)\left(\frac{t_0}{t}\right)^{(\gamma - 1)/(\gamma + 1)} - 2\right],$$

where t_0 is the time at which the particle path crosses the line $x = a_0 t$.

22. Use the jump conditions across a gas shock derived in §108 to show that in terms of the Mach number $M_0 = u_0/a_0$ *ahead* of the shock:

$$\rho_1/\rho_0 = u_0/u_1 = (\gamma + 1)M_0^2/\{(\gamma - 1)M_0^2 + 2\},$$

$$p_1/p_0 = \{2\gamma M_0^2/(\gamma + 1)\} - \{(\gamma - 1)/(\gamma + 1)\},$$

and that in terms of M_0, the Mach number M_1 *behind* the shock is

$$M_1^2 = \{2 + (\gamma - 1)M_0^2/\{2\gamma M_0^2 - (\gamma - 1)\}.$$

23. The equations of the long wave approximation for water waves can be expressed in the form

$$\frac{\partial h}{\partial t} + u\frac{\partial h}{\partial x} + h\frac{\partial u}{\partial x} = 0,$$

and

$$\frac{\partial u}{\partial t} + u\frac{\partial u}{\partial x} + g\frac{\partial h}{\partial x} = 0,$$

where g is the acceleration due to gravity, u is the horizontal component of the water velocity and h is the depth of the water. By re-writing the first equation in conservation form, and by multiplying each equation by a suitable factor and adding to form another equation which can also be expressed in conservation form, show that across a bore moving with speed $\tilde{\lambda}$:

$$\tilde{\lambda}[\![h]\!] = [\![uh]\!],$$

and

$$\tilde{\lambda}[\![uh]\!] = [\![u^2 h + \tfrac{1}{2}gh^2]\!].$$

Hence show that if side 0 of the discontinuity lies to the left of side 1, and the bore moves into water of depth h_0 which is at rest, the speed of propagation of the bore is

$$\tilde{\lambda} = -\sqrt{\left[g\frac{h_1}{h_0}\left(\frac{h_0 + h_1}{2}\right)\right]},$$

where h_1 is the water depth behind the bore, and that the water speed u_1 behind the bore is

$$u_1 = \tilde{\lambda}\left(1 - \frac{h_0}{h_1}\right).$$

Answers

Chapter 1

(2) $2\pi/(l^2+m^2)^{1/2}$;

(3) $\lambda = 2\pi/(A^2+B^2+C^2)^{1/2}$, velocity $= \lambda D/2\pi$;

(7) $A \sin nx \exp(-cnt)$;

(8) $A \exp\{-n(x+ct)\}$;

(9) $A \sin px \sin cpt$;

(10) $A \exp(-p^2 t) \sin px$, $p = \pi/l, 2\pi/l, \ldots$;

(11) Show that $\xi = $ const., $\eta = $ const., $\zeta = $ const. form an orthogonal system of coordinates, and transform $\nabla^2\phi$ in terms of ξ, η, ζ. The result is $\phi = X(\xi)Y(\eta)Z(\zeta)T(t)$, where m, p and q are arbitrary constants, and

$$\frac{1}{\sinh\xi}\frac{d}{d\xi}\sinh\xi\frac{dX}{d\xi} - \frac{m^2}{\sinh^2\xi}X + p^2 \sinh^2\xi X = q^2 X,$$

$$\frac{1}{\sin\eta}\frac{d}{d\eta}\sin\eta\frac{dY}{d\eta} - \frac{m^2}{\sin^2\eta}Y + p^2 \sin^2\eta Y = -q^2 Y,$$

$$\frac{d^2 Z}{dz^2} = -m^2 Z, \quad \frac{d^2 T}{dt^2} = -\frac{c^2 p^2}{a^2}T.$$

Chapter 2

(1) $1\cdot5$ m sec^{-1};

(2) $\frac{1}{4}Fa^2k^2 \sin^2 kct, \frac{1}{4}Fa^2k^2 \cos^2 kct$;

(3) $1/16$;

(5) $y = \sum b_r \cos\frac{(r+\frac{1}{2})\pi x}{a}\sin\frac{(r+\frac{1}{2})\pi ct}{a}$, $b_r = (-1)^r 4a^3/(r+\frac{1}{2})^4\pi^4 c$;

(6) $8\rho a^5/15$;

(7) $y = a_r \cos\frac{r\pi x}{l}\cos\left(\frac{r\pi ct}{l}+\varepsilon_r\right)$;

(8) energy in rth normal mode $= \dfrac{27c^2a^2\rho}{4l\pi^2r^2} \sin^2 \dfrac{r\pi}{3}$; sum $= 3c^2a^2\rho/4l$;

(11) $2\pi/p$ where $c_1 \tan (pl/c_1) = -c_2 \tan (pl/c_2)$, $c_1^2 = F/\rho_1$, $c_2^2 = F/\rho_2$.

Chapter 3

(1) $(2, 2)$ and $(4, 1)$: in general $(2m, n)$ and $(2n, m)$;

(2) $T = \frac{1}{2}\pi\rho n^2 c^2 A^2 \sin^2 nct \displaystyle\int_0^a \{J_m(nr)\}^2 r\, dr$,

$V = \frac{1}{2}\pi\rho c^2 A^2 \cos^2 nct \displaystyle\int_0^a [n^2\{J'_m(nr)\}^2 + m^2\{J_m(nr)\}^2/r^2] r\, dr$,

which becomes, after integration by parts,

$V = \frac{1}{2}\pi\rho n^2 c^2 A^2 \cos^2 nct \displaystyle\int_0^a \{J_m(nr)\}^2 r\, dr$;

(3) 391 per sec.;

(4) $z = A \sin (2\pi x/a) \sin (3\pi y/a - \cos (\sqrt{13}\pi ct/a)$;

(5) $z = A \sin (m\pi x/a) \sin (n\pi y/b) \cos \pi pt$, $p^2\rho = m^2 T_1/a^2 + n^2 T_2/b^2$.

Chapter 4

(1) 2 km per sec.;

(4) $R = \left\{ \dfrac{\sqrt{(\lambda_1\rho_1)} - \sqrt{(\lambda_2\rho_2)}}{\sqrt{(\lambda_1\rho_1)} + \sqrt{(\lambda_2\rho_2)}} \right\}^2$;

(5) $\xi = A_r \sin \dfrac{(r+\frac{1}{2})\pi x}{l} \cos \left\{ \dfrac{(r+\frac{1}{2})\pi ct}{l} + \varepsilon_r \right\}$;

(7) 1·690 sec., 0·252 sec.;

(8) Period $= 2\pi/nc$ where $k^2 - 3k \cot nl + \cot^2 nl = 1$, $k = Mln/m$.

Chapter 5

(1) $\frac{1}{4}g\rho la_r^2 \cos^2 \left(\dfrac{r\pi ct}{l} + \varepsilon_r \right)$, $\frac{1}{4}g\rho la_r^2 \sin^2 \left(\dfrac{r\pi ct}{l} + \varepsilon_r \right)$;

(2) radial velocity $-(gA/c) \cos m\theta J'_m(nr) \sin (nct + \varepsilon)$, transverse velocity $(gAm/ncr)\sin m\theta J_m(nr) \sin (nct + \varepsilon)$, amplitude $(gA/c)J'_0(nr)$;

(3) $\zeta = A \cos (p\pi x/a) \cos (q\pi y/a)$ $(\cos r\pi ct/a)$, $r^2 = p^2 + q^2$ and kinetic energy $= \frac{1}{8}g\rho A^2 a^2 \sin^2 (r\pi ct/a)$, potential energy $= \frac{1}{8}g\rho A^2 a^2 \cos^2 (r\pi ct/a)$;

(4) $\dfrac{X}{Y} = \dfrac{p}{q} \tan \dfrac{p\pi x}{a} \cot \dfrac{q\pi y}{a}$;

(8) $X:Y:Z = nrJ'_m(nr):-mJ_m(nr)\tan m\theta: nrJ_m(nr)$

(9) Same as in equations (39) to (41) except that $1-m = 5k/2$, where $k = 0, 1, 2, \ldots$.

Chapter 6

(2) Reflection coefficient $R = \{1 + 4\rho_0^2/M^2 n^2\}^{-1}$,

transmission coefficient $T = \{1 + M^2 n^2/4\rho_0^2\}^{-1}$;

(5) Period $= 2\pi/pc$, where $(abp^2 + 1)\sin p(b-a) = p(b-a)\cos p(b-a)$;

(7) 166 per sec.

Chapter 7

(2) $H_x = H_z = 0$, $H_y = -A \sin nx \sin nct$;

(4) $9°\ 28'$, $9°\ 36'$;

(7) 12/13 of the incident energy is transmitted;

(8) $100°\ 20'$;

Chapter 8

(1) 33·2 km per sec., 249 per sec.;

(3) $a + c/2, b/2, c/4$;

(5) $V\left\{\dfrac{A\lambda_0^2\lambda^2}{(\lambda^2 - \lambda_0^2)(\lambda^2 - \lambda_0^2 + A\lambda^2)}\right\}$, $V\left\{1 + \dfrac{A\lambda^2(B\lambda^4 - \lambda_0^2(\lambda^2 - \lambda_0^2))^2}{\varkappa_e\{(\lambda^2 - \lambda_0^2)^2 + B\lambda^2\}^2}\right\}$

where V = wave velocity;

(6) $U = \frac{1}{2}c + 2\pi T/\lambda\rho c$.

Chapter 9

(1) One possibility would be to set $u = \partial y/\partial t + \partial y/\partial x$ and $v = \partial y/\partial t - \partial y/\partial x$, there are many others;

(2) $\dfrac{\partial \mathbf{U}}{\partial t} + \mathbf{A}_1 \dfrac{\partial \mathbf{U}}{\partial x} + \mathbf{A}_2 \dfrac{\partial \mathbf{U}}{\partial y} + \mathbf{A}_3 \dfrac{\partial \mathbf{U}}{\partial z} = \mathbf{0}$

with

$$\mathbf{U} = \begin{bmatrix} \rho \\ u \\ v \\ w \end{bmatrix}, \quad \mathbf{A}_1 = \begin{bmatrix} u & \rho & 0 & 0 \\ a^2/\rho & u & 0 & 0 \\ 0 & 0 & u & 0 \\ 0 & 0 & 0 & u \end{bmatrix}, \quad \mathbf{A}_2 = \begin{bmatrix} v & 0 & \rho & 0 \\ 0 & v & 0 & 0 \\ a^2/\rho & 0 & v & 0 \\ 0 & 0 & 0 & v \end{bmatrix},$$

$$\mathbf{A}_3 = \begin{bmatrix} w & 0 & 0 & \rho \\ 0 & w & 0 & 0 \\ 0 & 0 & w & 0 \\ a^2/\rho & 0 & 0 & w \end{bmatrix}.$$

(3) $\dfrac{\partial \mathbf{U}}{\partial t} + \mathbf{A}\dfrac{\partial \mathbf{U}}{\partial x} = \mathbf{0}$

(with $\mathbf{E} = (E_2, E_3)$, $\mathbf{H} = (H_2, H_3)$ and

$$\mathbf{U} = \begin{bmatrix} E_2 \\ E_3 \\ H_2 \\ H_3 \end{bmatrix}, \quad \mathbf{A} = \begin{bmatrix} 0 & 0 & 0 & -1/\varepsilon_0\varkappa_e \\ 0 & 0 & 1/\varepsilon_0\varkappa_e & 0 \\ 0 & 1/\mu_0\varkappa_m & 0 & 0 \\ -1/\mu_0\varkappa_m & 0 & 0 & 0 \end{bmatrix};$$

(4) $\dfrac{\partial \mathbf{U}}{\partial t} + \dfrac{\partial \mathbf{F}}{\partial x} = \mathbf{0}$

with $\mathbf{U}$ defined as above and

$$\mathbf{F} = \begin{bmatrix} -H_3/\varepsilon_0\varkappa_e \\ H_2/\varepsilon_0\varkappa_e \\ E_3/\mu_0\varkappa_m \\ -E_2/\mu_0\varkappa_m \end{bmatrix};$$

(5) $\dfrac{\partial \mathbf{U}}{\partial t} + \dfrac{\partial \mathbf{F}}{\partial x} = \mathbf{0}$

with

$$\mathbf{U} = \begin{bmatrix} u \\ c^2 \end{bmatrix}, \quad \mathbf{F} = \begin{bmatrix} \frac{1}{2}u^2 + c^2 - H \\ uc^2 \end{bmatrix};$$

(7) Characteristic curves: $x = \xi + t \sin \xi$ through point $(\xi, 0)$ of initial line.
Implicit solution: $u = \sin(x - ut)$
Envelope in parametric form: $x = \xi + t \sin \xi$, $\qquad t = -1/\cos \xi$.

Index